Joseph Henry Wythe

The Science of Life

Or the Animal and Vegetable Biology

Joseph Henry Wythe

The Science of Life
Or the Animal and Vegetable Biology

ISBN/EAN: 9783337034276

Printed in Europe, USA, Canada, Australia, Japan

Cover: Foto ©berggeist007 / pixelio.de

More available books at **www.hansebooks.com**

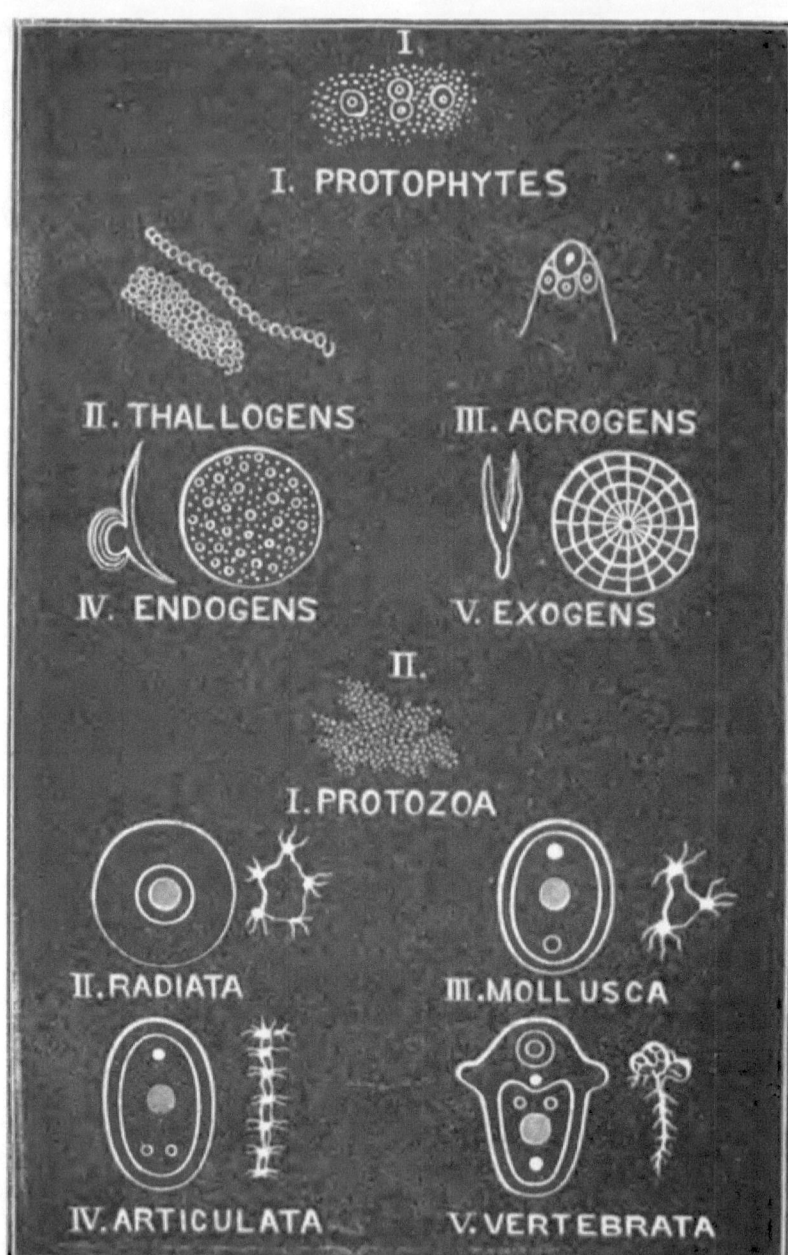

I

I. PROTOPHYTES

II. THALLOGENS   III. ACROGENS

IV. ENDOGENS   V. EXOGENS

II.

I. PROTOZOA

II. RADIATA   III. MOLLUSCA

IV. ARTICULATA   V. VERTEBRATA

BIOLOGICAL TYPES.

THE

SCIENCE OF LIFE;

OR,

ANIMAL AND VEGETABLE BIOLOGY.

BY

REV. J. H. WYTHE, A.M., M.D.,

AUTHOR OF "AGREEMENT OF SCIENCE AND REVELATION," "THE MICROSCOPIST," ETC

———— ‹•─•·•─ ————

NEW YORK:
PHILLIPS & HUNT.
CINCINNATI:
WALDEN & STOWE.
1880.

PREFACE.

THIS book is written for those who have some elementary knowledge of Physiology. It gives a general outline of the origin, structure, typical forms, and functions of living things, so as to serve as an introduction to the examination of the objects themselves.

Although a text-book must of necessity be a compilation of facts, yet many years of practical experience with the microscope have enabled the writer to describe many things with the confidence of personal observation. Some of the illustrations are original, others have been selected from Dr. Carpenter's works on Physiology and the Microscope, T. R. Jones on Zoology, Lindley's Botany, Mac Ginley's Introduction to Biology, and other standard works.

It has been the aim of the author to guide the student through the fundamental principles of

1*

Biology to the contemplation of the vast temple of animated nature, with its varied compartments intimately connected with each other, and with the central one of all, the human type. In every avenue and chamber and dome of this wondrous edifice the Christian student recognizes the truth that " POWER BELONGETH UNTO GOD."

OAKLAND, CAL., *January*, 1880.

CONTENTS.

SCIENCE OF LIFE.

CHAPTER I.

WHAT IS LIFE?

Am I but what I seem—mere flesh and blood?
A branching channel, and a mazy flood?
The purple stream that through my vessels glides,
Dull and unconscious flows, like common tides.
The pipes, through which the circling juices stray,
Are not that thinking I, no more than they.
This frame, comparted with transcendent skill,
Of moving joints, obedient to my will,
Nursed from the fruitful glebe, like yonder tree,
Waxes and wastes: I call it *mine*, not *me*.
New matter still the moldering mass sustains,
The mansion changed, the tenant still remains;
And from the fleeting stream, repaired by food,
Distinct, as is the swimmer from the flood.

—ARBUTHNOT.

1. THE term Biology, (from the Greek, *bios*, life, and *logos*, a discourse, or doctrine,) signifies the Science of Life. It includes the study of all the phenomena of living beings, both animal and vegetable, in order to discover the general principles which underlie their origin, formation, varieties, and functions. The special study of structure is termed Morphology, or Anatomy. The study of functions is Physiology. The origin, development, and arrangement of the varieties of the vegetable world make up the study of Botany. Zoology considers

the various kinds of animals. All these sciences, and many others, combine in Biology.

To the Christian student Biology affords a multitude of evidences of intelligent design, proving the universe to be the product of Supreme Will. It also contains proof of the reality of spiritual existences, in addition to physical atoms and physical forces.

2. The cause of difference between the living and the non-living is the most fundamental question of Biology, and the answers given to this question by modern writers depend upon the schools of philosophy to which they are attached.

Much learning and industry have been employed within the past few years to teach the system of *Monism*, or the theory that all being can be resolved into a single principle. Among those who entertain this view, some hold to materialism, or the development of all forms from primitive atoms. Others are idealists, conceiving matter to be identical with force. Others again are pantheists, holding that mind is the only substance, and that the universe is an emanation of the universal mind.

The doctrine of rational *Dualism*, which asserts two real principles of existence, mind and matter, with their special endowments and forces, stands in opposition to all forms of Monism whatever.

Since the dawn of history these speculations have divided philosophers, and learning of all kinds has been used to maintain the views of either side. Leucippus and Democritus, the masters of Epicurus, taught the doctrine of invisible and indestructible atoms, with spontaneous motion, as the cause of all things. Anaxagoras and Plato argued for a regulating intelligence, producing order, so that "the world's activities are reflections of God's thoughts." The Hebrew and Christian Script- ures, as well as all other writings which exhibit the religious beliefs of mankind, Koran or Shaster, King or Avesta, (the sacred books of Mohammedans and Hindus, Chinese and Persians,) teach the doctrine of Dualism, or the distinction between mind and matter.

3. The revival of Monistic philosophy in the last century has awakened much discussion, and each of the sciences in turn has been made the arena of conflict. In Biology, Darwin, Spencer, and Hæckel are arrayed against Agassiz, Lionel Beale, and M'Cosh, and the con- test of mind has brought to notice a wonderful accumu- lation of facts, sufficient, we think, to settle the central question of philosophy concerning life.

In the present work the facts of Biology are regarded as confirmatory of the principles of rational Dualism. In the judgment of the writer there is no conflict between science and revealed truth, but such complete agreement that the facts of science can be best understood and explained in consistency with that philosophy which re- ligion has made prevalent in the minds of the majority of men. Yet the learning and apparent candor of many Monistic writers entitle them to respect, even if we fail to agree with them, and truth, which should be the object of all study, is not aided by epithets or personal acri- mony.

4. Some scientists ignore the question of the cause of life, and confine themselves to the physical and chemical phenomena associated with living things; but this is

quite unsatisfactory. That there are differences between the living and the non-living will only be denied by the most thorough partisans of Monism. These differences depend on something in the living which is absent from the non-living. In common parlance we call it life, or life-force. Such a life-force is as necessary to Biology as gravitation is to Physics, or light to Optics.

Writers who avoid Dualism, or who acknowledge antagonism to it, have not been able to give a clear definition of life.

Bichat defines life as "the sum of the functions by which death is resisted." This is but saying that life and death are opposite states.

Dr. W. B. Carpenter, although believing in the difference between mind and matter, speaks of life as "the condition of a being which exhibits vital actions;" which is but another mode of stating that life is a condition or state of living.

Coleridge considered life as synonymous with "individuation." This is equivalent to separate existence, and includes metals, and stones, and all non-living things.

Herbert Spencer defines life as "the continuous adjustment of internal relations to external relations." This definition will apply to a boiling tea-kettle, a steam-engine, or a burning candle, as well as to a living thing.

Haeckel declares "that all natural bodies which are known to us are equally animated, and that the distinction which has been made between animate and inanimate bodies does not exist." This exceedingly bold and strange statement is rendered necessary by the logical demands of the Monistic philosophy. In a subsequent place we shall examine particularly the differences between animate and inanimate bodies. (See Chap. II.)

All such definitions and statements evade the real question: that is, What makes the difference between a living body and the same body a moment after death?

5. The cause of life is a mystery only to the materialist. To the Christian philosopher it is as plainly revealed as any other fact of nature. The Bible asserts that life results from the union of a spiritual nature with the material body. In other words, life is the influence resulting from the union of matter and spirit; and this dualistic theory is the only one which suffices to explain the phenomena of living things.

Moses declares of man that God " breathed into his nostrils the breath of life; and man became a living soul." In accordance with this view death is everywhere referred to in Scripture as a departure of the spirit. The medical evangelist, St. Luke, when describing the resuscitation of Jairus' daughter, says, " Her spirit came again, and she arose straightway." St. Paul describes the body as a tent, or house, in which the spirit may be present or absent. It is also remarkable that the same Hebrew word which describes man as a "living soul" is applied to animals in the same history of creation. Gen. i, 20, 30. They also are living souls.

This view of the cause of life was also held by ancient Grecian philosophy. Aristotle attributed organization and vital actions to a series of *animating principles*, (*psychai*,) different in each organized body, and acting by power derived from the supreme animating principle, (*physis*.)

2

Müller, the father of modern physiology, substituted the term "*organic force*" for that of "animating principle," and Dr. Prout used the term "organic agent." The precise term employed is of but secondary importance compared with the dualistic conception, which is quite satisfactory to the large majority of thinkers.

6. We shall be able to appreciate this subject better if we consider the life-history of some simple animal.

It is well known that infusions of vegetable or animal substances contain many living forms of extreme simplicity of structure, called *Infusoria*. Many such are found in ponds, or running water, or in the sea. A very beautiful kind of Infusoria, common among half-decayed leaves, has received the name of Vorticella, or bell-shaped animalcule. There are several species, the most common being known as *Vorticella nebulifera*. Take up from a pond a little twig, covered with mold or mucus-like substance, and place it under the microscope. In all probability you will see a colony of Vorticellæ, (Fig. 1.)

Each animalcule has a glassy, transparent bell, with a thick lip or rim, fringed with cilia or hair-like projections. These cilia are sometimes withdrawn, but when active vibrate rapidly, so as to make a sort of whirlpool in the water, in the vortex of which smaller animals or vegetables may be conveyed as food to the interior of the Vorticella. A number of pellucid spots may be seen in the body of each animalcule, which were formerly regarded as stomachs. Professor Ehrenberg, who elaborately investigated this class of animal life, gave the name *Polygastrica* (many stomachs) to those animal-

cules which presented this appearance. By feeding with
coloring matter, as carmine or indigo, these stomachs
have been found to be merely excavations in the bioplasm, or living matter, which constitutes the body of

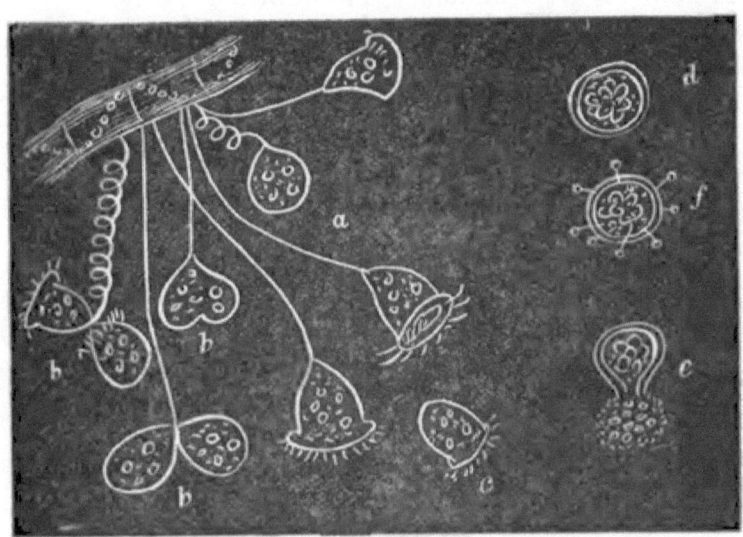

FIG. 1.—*a.* Colony of Vorticella. *b, b, b.* Stages of fission, or self-division. *c.* A separate individual. *d.* Encysted state. *e.* Ruptured cyst emitting gemmules in a mass of
gelatine or gum. *f.* Acineta parasites.

the animal. Some of these excavations are extempo-
raneous, but one cavity is persistent, and pulsates in a
peculiar manner, so that it has received the name of
contractile vesicle. Each glassy bell is attached to the
twig by a slender thread, and usually swings to and fro
in the water with the thread or footstalk fully stretched,
and the cilia moving rapidly. Frequently, however, and
especially on some unusual jar, or other cause of alarm,
the thread contracts in the form of a spiral, and the cilia
are withdrawn into the substance of the bell.

These Infusoria usually increase by self-division. The

globular bell becomes first flattened, then notched, and lastly divided. As soon as division takes place there are distinct motions in the separate individuals. In one of them the cilia are absorbed, and new cilia appear on the side next to the footstalk. The motions of the new cilia form a current sufficient to detach the newly-formed bell, which becomes isolated, swims away, and develops a new stalk, after fixing itself in a new place.

Another mode of increase sometimes occurs, in which the animalcule seems to pass through a sort of chrysalis state. It becomes encysted, like the primitive forms of vegetables. It is first rounded, then a sort of gelatinous secretion hardens into a case, protecting the interior from antagonizing cold, etc.; then the encysted body breaks up into nuclei, or separate spots, and afterward into numerous gemmules, or small germs, which are set free by the bursting of the envelope, and swim away to grow into new individuals.

During the encysting process the Vorticella often appears like a globular pincushion with pins sticking in it. This is now known to be caused by a parasite, the *Acineta*, which sends forth a projecting arm into the body of its host to absorb its fluid nutriment.

7. I have selected the Vorticella for a first lesson on Biology because it is quite common, and simple enough for study. What can we learn here of life-force? Is there such a thing as life-force? Is there a difference between the living Vorticella and the dead twig it rests upon? Some philosophers, as we have seen, declare that there is no difference. The old astrologers used to say that all things were living, and the teachers of an-

cient magic and heathen philosophy taught a universal world-spirit, which is the life of all things. To this pantheistic theory the adherents of the dogma of the mechanical origin of the universe naturally gravitate. It is more consistent with common sense and true philosophy, as well as with the facts of science, to maintain an essential difference between the animate and the inanimate. Can the dead twig move spontaneously, like the living animalcule? Does it assimilate food and reproduce itself like the Vorticella? Or can a dead animal respond to natural stimuli like the living? Not a single fact has been brought forward to prove the identity of the living and the non-living. It is at best only a theory. "On the other hand," says Dr. Beale, "thanks to the steady progress of minute investigation, unnoticed by popular writers, and perhaps unknown to them, the conclusion that life of every kind is distinct from ordinary forces is at this time more strongly supported by facts, and more firmly established than it ever was." *

8. In order to defend the Monistic philosophy, and the identity of animate and inanimate objects, some argue that matter has no existence as such, but that each atom is only a center of force. They thus repudiate the charge of materialism, since they teach that every thing is spirit. This is a most subtle and ingenious method of defense, yet is just as baseless as the grosser Monism, which considers all to be material.

Newton's law, of gravity being in direct ratio to the mass of matter, that is, to the number of atoms in the

* Beale's "Protoplasm."

2*

mass, proves atoms to be real physical existences. All chemical science is based on the doctrine that atoms and molecules have weight, definite proportions or relations, and hence definite form. The law of Avogadro and Ampere, as it is called, that "equal volumes of all substances when in the state of gas, and under like conditions, contain the same number of molecules," is confirmed by all chemical experiments, and necessarily implies the reality of atoms and molecules. Our own consciousness of matter, also, the sense of otherness which pertains to our knowledge of the objects of sense, is as reliable as any other knowledge. We know the *otherness*, as well as the weight and inertia of matter by the same faculties by which we know that two and two make four, and not five. The obvious distinctions between the living and the not living are all proofs of Dualism.

9. As to the theory that atoms have a physical and a spiritual side, by which opposite qualities are exhibited, it carries its own refutation, since it is plainly impossible for a healthy mind to believe that contrary properties can inhere in any thing at the same time. Mr. Joseph Cook has pertinently said: "If matter is a double-faced unity, having a spiritual and a physical side, there must co-inhere in one and the same substratum extension and the absence of extension, inertia and the absence of inertia, color and the absence of color, form and the absence of form. To assert that these fundamentally antagonistic qualities of matter and mind not only inhere, but co-inhere, in one and the same substratum, is to assert that a thing can be and not be at the same time and in the same sense. This limitless self-contradiction wrecks

in this age, as it has wrecked in every age, the pretense
that there is but one substance in the universe.*

10. The continuance of life in an organism composed
of new atoms, after the old atoms have been cast off,
proves that the cause of life does not spring from the
atoms themselves. An atom of oxygen or hydrogen,
endowed with life to-day, as part of an organized molecule of a Vorticella, or as part of our own bodies, may
be to-morrow released from its vital connections, and be
transported, as water or air, to remote parts of the globe.
It may form part of the gigantic Sequoias of the Sierras,
the Cinchona-trees of the Andes, or the Rhododendrons
of the Himalayas. Before the death of the original
organism, or the tree it next served, that atom of oxygen or hydrogen may be again discarded, and pass into
the germ-cell of an animal, or become part of one of the
tissues of a man in a distant part of the world. It is
evident that *that* atom did not produce the life with
which it was first associated. What may happen to one
atom may happen to all the atoms of an organism. In
active living beings this actually does happen, so that all
the atoms of a living body become disconnected, and
return to the inorganic world, or go to serve other organisms, while other atoms take their places, yet the
organized body lives on. Its life depends not on the
new atoms, for the body was animate before these atoms
came; nor does it depend on the old atoms, for it continues after they have gone. It must, therefore, depend
upon something different from the material atoms. As
matter and spirit are the only objects of thought pos-

* Cook's "Biology," p. 227.

sible to us, and as life does not depend on matter, it must depend on spirit. If existence and activity continue after the removal of the original matter, as we have seen, they may also continue after all matter is removed. Continued spiritual existence is certainly conceivable, and in view of the endowment of new atoms by the vitalizing force, we must admit it to be probable, even after the material of the organism is all destroyed.

The cause of life is more than matter and physical force. It uses both matter and force for its own ends and after its own laws. "Its power of control over matter and physical laws proves its superiority over, and its distinction from, matter. Life is matter's master, not its slave. Life is a workman, a builder, a chemist; and each organized being has its own appropriate life, the result of the union of the spiritual and the material in itself." *

11. The view we have taken of the difference between the animate and the inanimate objects of creation is one which is growing in favor with the principal workers in biological science. Dr. Beale's discoveries and generalizations in Histology have done much to arrest the skeptical tendencies of scientists, and in one of Mr. Huxley's latest utterances he acknowledges that "the properties of living matter distinguish it absolutely from all other kinds of things," and that "the present state of knowledge furnishes us with no link between the living and the not-living." † The last-named anatomist names the distinctive properties of living matter as fol-

* "Agreement of Science and Revelation," by the Author.
† Huxley's "Anatomy of Invertebrated Animals."

lows: 1. Its chemical composition; 2. Its universal disintegration and waste by oxidation, and its concomitant
reintegration by the intussusception of new matter;
3. Its tendency to undergo cyclical changes.

Dr. Beale shows that "no relation can be established
between the chemical or other material properties of
different kinds of living matter that will in any way account for the different results as regards development
and formation. The different powers or properties of
the particles cannot be due to difference of chemical
composition. All living particles consist of comparatively few elements, and no differences in the proportions of these would enable us to explain the different
results of the act of living.

"This wonderful stuff, which is the first state of every
thing that has life, splits up when it is destroyed into
a few chemical compounds, from the study of which,
however, chemists have hitherto failed to arrive at any
conclusion as regards the atomic relations of the component elements of the matter during life. Neither, as
far as has been ascertained, is there any constant relation between the volume, or kind, or aggregation of the
matter which is the seat of the manifestation of the vital
power and the form of living being that is to be evolved
from it. Man's matter is no more elaborate, no more
complex, no more beautiful, than dog's matter or sheep's
matter; but it is in the *power*, not in the matter, that
we must look for the cause of the remarkable difference
of the results. Insignificantly in matter, but transcend-
ently in power, does the man-germ differ from the dog-
germ. Wonderfully different power may be transmitted

by particles of matter that resemble one another in every particular that can be ascertained." Again: "It is by the transmission of power to matter, rather than by the bodily transference of millions of particles of matter having particular properties and detached from matter having similar properties, that inheritable peculiarities are handed down from parent to offspring. And it must be borne in mind that structure-forming capacity, which is not even rendered evident until forty or fifty years shall have passed since the original germ-speck originated in the parent, may affect pounds weight of matter, not one grain of which will be acquired until long after every atom of that primitive speck shall have ceased to live and have been removed from the organism. Matter, with its forces, continually comes and goes, while power only remains unimpaired and preserves its identity. Power has been handed down—has been transferred from old particles to new particles of matter; but the original matter—nay, in the case of some of the largest animals, hundreds weight of matter—must have come and gone, while the original power remained." "Vital power works according to predetermined order, and the results of its working are seen in different consequences, at different periods of its action." "Vital power prepares for far-off events, and acts as if phenomena, not to occur until after the lapse of a considerable time, had been from the first foreseen. Vital power suspends the action of chemical affinity, and piles material particle above particle, the force of gravity notwithstanding."*

12. Sometimes life remains *dormant* from lack of ap-

* "Protoplasm," by Dr. L. Beale.

propriate stimuli, or conditions, or from some unexplained peculiarity. This proves those philosophers to be in error who imagine that molecular change is essential to life. The seed which has been held in the hand of an Egyptian mummy perhaps for thousands of years, retains the vital power, and may sprout under favorable conditions. The wheel animalcule (Rotatoria) has been dried and resuscitated many times in succession, and Messrs. Drysdale and Dollinger have proved that the germs of Infusoria cannot be destroyed by the heat of boiling water, but live when the thermometer shows a heat of 300° F. These resisting germs, floating in the air, will soon revive on the accession of moisture.

13. *Death* occurs when the cause of life is removed. Life is not synonymous with spirit, but is peculiar spiritual influence on matter; the result of the union of created spirits and elemental matter. When the spiritual essence ceases to act upon the matter of the organism we say the body is dead, and then disintegration and chemical decomposition succeed. There is a two-fold death—the death of the organism as a whole, called somatic, or bodily death, and molecular death, or the loss of vital activity in the molecules of the body. Life begins in a single molecule of bioplasm, and is propagated as a force more or less modified from molecule to molecule, or from cell to cell, as flame proceeds from one combustible substance to another, or as magnetism is disseminated by the action of a single magnet through one bar of steel after another.

Molecular death is a continual phenomenon of life during its activity. It is arrested in dormant life, and

is far from being so constant an attendant upon all the
actions of the body as some have taught, yet it goes on
with great rapidity and uniformity. The bioplasts, or
living particles, of each tissue in the body are changed
into formed material, and then pass into decay, while
other bioplasts take their places and keep up the active
dance of life. When the spiritual cause, or origin, of
vital phenomena is removed, the molecular activities of
the body do not all cease at once, but gradually. Hair
will continue to grow on a corpse, and the secretion of
rattle-snake poison, or of other glands, continues for a
short time after death. Indeed, the circulation of blood
has been witnessed in a section of mouse's kidney some
time after it had been removed from the body. Yet,
uninfluenced by the energizing spirit, the vital activities
gradually cease, and decomposition ensues.

14. To return to our example from the Infusoria, the
life-history of the Vorticella demonstrates both the spiritual origin of life and the work of a Supreme Intelligence. The evidence of design in its construction is
quite apparent. The extensile threads and vibsatile
cilia have, plainly enough, an object. They subserve
prehension of food and the preservation of existence.
Even the contractile vesicle, whose exact purpose we do
not know, impresses our minds with the fact that it serves
some purpose. This design is connected with something different from the material atoms of the organism,
but which controls those atoms, since there is foresight
of future changes, and provision for future changes in
the life-history which will occur after the removal of all
the present material. The self-division of the Vorticella,

the formation of new cilia, the preparation for increase
by the encysted form, the division into nuclei and gemmules, are all examples of this, analogous to the formation of new structures in the higher animals. The power
to produce these changes is not material but spiritual.

15. Thus our first lesson in Biology brings us to the
confines of a spiritual world. We look across the gulf
which philosophy and science cannot bridge over except
by revealed truth, but the telescope of faith can see re-
alities on the other side as numerous, as diversified, and
as true as the objects of sense which can be weighed and
measured by our physical instruments. We see also the
care and providence of a Supreme Creator. Astronomy
adds emphasis to the Psalmist's declaration, "The heav-
ens declare the glory of God; and the firmament showeth
his handy-work." And Biology indorses the sentiments
of his eloquent utterances respecting living creatures:
"O Lord, how manifold are thy works! in wisdom hast
thou made them all: the earth is full of thy riches. So
is this great and wide sea, wherein are things creeping innumerable, both small and great beasts. There
go the ships: there is that leviathan, whom thou hast
made to play therein. These all wait upon thee; that
thou mayest give them their meat in due season. That
thou givest them they gather: thou openest thine hand,
they are filled with good. Thou hidest thy face, they are
troubled: thou takest away their breath, they die, and
return to their dust. Thou sendest forth thy spirit, they
are created: and thou renewest the face of the earth.
The glory of the Lord shall endure forever: the Lord
shall rejoice in his works."

3

CHAPTER II.

LIVING MATTER.

You may bury me as you choose, if you can only catch me. But you will not understand me when I tell you that I, Socrates, who am now speaking, shall not remain with you after having drunk the poison, but shall depart to some of the enjoyments of the blest. You must not talk about burying or burning Socrates, as if I were suffering some terrible operation. Such language is inauspicious and depressing to our minds. Keep up your courage, and talk only of burying the body of Socrates ; conduct the burial as you think best and most decent.—PLATO'S Phædo.

1. THE only unexceptionable characteristic of living bodies is the possession of living tissue, or *bioplasm*. This may be present alone, as in the simple animal and vegetable forms, or it may exist in association with structure which has been formed by it, and hence called *formed material*. The bioplasm is nourished by *pabulum* which is generally furnished in fluid form.

2. The old division of bodies into organized and un-

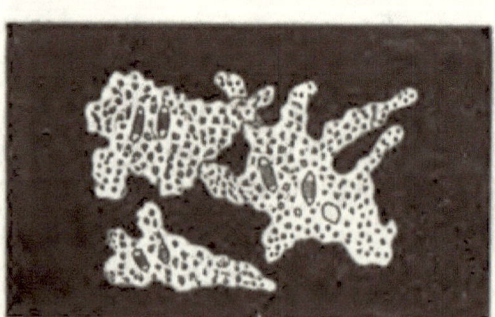

FIG. 2.—Amœba princeps × 150. In various shapes.

organized—the former having organs, or distinct parts, with definite structure, and of special use — is no longer applicable, since there are some living things which have no organs. The *Amœba princeps*, Fig. 2, one of the most elementary animal forms, is composed of a jelly-

like homogeneous bioplasm, capable of indefinite exten-
sibility and of indefinite use. It is so constantly alter-
ing its outline that it does not retain the same shape for
two successive minutes. It obtains its food by flowing
around it, and digests by direct absorption.

3. Of such simplicity of structure are all the primitive
forms of vegetable and of animal life, while in bone, cartilage, flesh, skin, or any other structure of the higher
animals, we find such simple, jelly-like, living matter, or
bioplasm, similar in appearance to the Amœba, scattered in minute particles all through the tissue, and
careful observation will show how this living matter
is transformed into the formed material of the several
tissues.

4. All animals and vegetables have originated from
minute particles of such bioplasm. Every dog, horse,
man, whale, jelly-fish, oak, cedar, grass, sea-weed, etc.,
began its existence as a particle of bioplasm. And
every tissue and organ, no matter what its form or function, was built up by similar living matter.

5. In the lowest type of animal life (the Rhizopods)
the vital operations are carried on without any special
organs, as we have seen in the Amœba; a little particle
of jelly-like bioplasm, changing itself into a variety of
forms, laying hold of food without members, swallowing
it without a mouth, digesting it without a stomach,
moving without muscles, while the mere separation of a
fragment of this jelly, however small, is sufficient to
originate another and independent living creature, retaining, or rather repeating, all the characteristic endowments of the original mass. In the higher animals,

although the first bioplasmic particle subdivides itself into an aggregation of similar particles or cells, yet there soon appears a structural differentiation of organs for special uses, which is more elaborate and heterogeneous as the type approaches the human structure. A single cell or living particle, however, in any structure is, to all intents and purposes, a living thing, and possesses powers of assimilation, growth, and reproduction, altogether different from the mineral or non-living body.

6. Living matter, or bioplasm, may be considered physically as a peculiar compound of the chemical elements — carbon, oxygen, nitrogen, and hydrogen, called by Mulder *Proteine*, and by Mr. Huxley and the German histologists *Protoplasm*, or the physical basis of life. It is nearly identical with Albumen. So far as is known, this combination of elements is always the product of pre-existing, living matter. It has never been produced in the laboratory, and if it were possible for a chemist to manufacture albuminoid matter, or protoplasm, it would be dead protoplasm, and not bioplasm, and would be destitute of vital properties. Other conditions are necessary to vital phenomena besides combination of material elements. Light, heat, electricity, and moisture are all necessary conditions; nor these alone, for with all these existing and active, the protoplasm may not live. Some other factor is essential to life besides matter and physical force, as we said in the last chapter. The term *bioplasm* is well applied to express matter in its living state, while *protoplasm* should be restricted to the material itself.

7. The essential phenomena of living matter next

claim our attention; or, What can a living thing do which the non-living cannot?

1.) All living things have *spontaneous motion*. The non-living are passive, and only move by the compelling agency of some external force, but the force which moves living matter is a force which is inherent, and cannot be explained by physical laws. Living matter has primary energy, and can overcome inertia, but the non-living are unable to originate motion. The spontaneous motions of bioplasm, or living matter, are molecular, amœboid, or wandering.

a. Molecular movement. This must not be confounded with what has been called Brunonian motion, from Dr. Robert Brown, who first described it in 1827. The latter is a sort of vibration in small particles suspended in fluid, and is supposed to be caused by currents formed by heat or evaporation. In the molecular movements of bioplasm each particle of the mass seems to be independent of the rest. As the passengers in a crowded street may go the full length of the street, or turn back, or stop and double as many times as they wish, so do the particles move in the mass of bioplasm. Up, down, across, backward, and in all directions—even through each other—do these molecules move, each impelled by its own inherent energy.*

b. Amœboid movement receives this name from its resemblance to the notions of the Amœba, described in the present chapter, Sec. 2. The shape is continually changing, by a portion of the body being projected from the mass, or retracted, or altered in form.

* Stricker's "Manual of Histology."

c. Wandering movement is a modification of the latter form. A portion of the bioplasm is projected forward, and along this temporary arm, or bridge, the semi-fluid molecules flow along, and accumulate at the farthest end. In this manner the white cells of blood, which are particles of bioplasm, wander out of the vessels, perhaps by means of stomata, or holes, in the sides of the vessels, into those tissues of the body where they are needed,

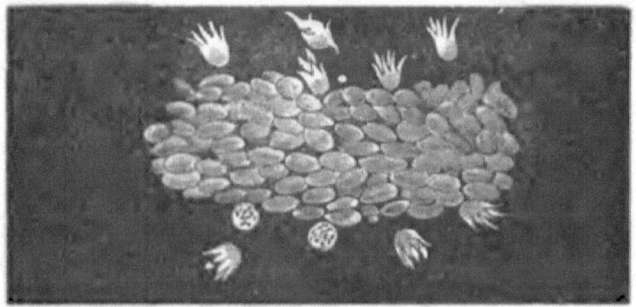

FIG. 3.—Clot of Frog's Blood, with Migrating White Blood-cells.

Fig. 3. These motions are wholly unlike any which occur in lifeless material.

2.) Another essential property of bioplasm is *growth.* The term growth does not mean accretion or addition of material, nor increase of size. A piece of chalk, or a bank of mud, or any non-living thing, may increase in size by additions to its material. Growth in a living thing is different. It is enlargement by nutrition, and depends on inherent motion. In Chap. I, Sec. 13, it was stated that hair would grow on a corpse, but the term grow was used in a popular, and not scientific, sense. Hair is not a living part of the body. Hair or nails may be cut or destroyed without sensation or impairment of the body. They consist of scales of formed material, pushed forward by the growth of bioplasm behind them. If you pull out a hair or nail, you reach the quick—that is, the living or sensitive part. We thus see that some parts of our body are alive, and others in a non-living state. The formed portions never grow, but the bioplasm, or living matter, grows. The growth of living matter is by appropriation and transformation. Bioplasm "alone, of all matter in the world, moves toward lifeless matter, incorporates it with itself, and communicates to it, in some way we do not in the least understand, its own transcendentally wonderful properties." This motion and incorporation and endowment constitute growth.

"The rootlets of the plant extend themselves into the soil because the living matter at their extremities moves onward from the point already reached. The tree grows upward against gravity by virtue of the same living power of bioplasm. In every bud portions of this living matter tend to move away from the spot where they were produced, and stretch upward and onward in advance. No tissue of any living animal could be formed unless the portions of bioplasm moved away from one another." *

3.) Living matter has also the power of nutrition, or assimilation by selection. As this is connected with growth, we might have considered it under that head, but since writers of the mechanical or materialistic school attempt to account for it on physical or chemical principles, we deem it best to examine it separately.

The non-living always enlarges by accretion from sim-

* Beale's " Bioplasm."

ilar material; the living tissue takes into its interior material which it transforms out of pabulum, which is foreign
to its own structure, while at the same time it discards
such molecules or atoms as are unfit for further use.

The chemical composition of the various tissues of the
body cannot be found in the blood, or pabulum, which
nourishes the tissues, but results from metamorphosis, or
transformation, by means of the bioplasts. Endosmose,
or the physical property by which fluids pass through
membranes, or gummy matters, will not account for it,
since in the latter there is no change of material, while
in nutrition there is rearrangement of the atoms in the
tissue-molecules.

Nutrition has sometimes been compared with crystallization, but crystallization is a deposit of material from a
solution of similar substance, and is altogether different
from nutrition by transformation and selection.

Nutrition has also been compared with a chemical
phenomenon called catalysis. In this, chemical change
takes place because of the presence of a substance which
remains itself unaffected, as when spongy platinum in-
duces the combination of oxygen and hydrogen gases.
In catalysis the third substance neither gives nor takes
from the excited body, but in nutrition the living matter
itself selects appropriate chemical elements from its pabulum, dissolving their former affinities, and recombining
them in a manner which no non-living substance can do.
There is no third substance present which is known to
us, and all the phenomena are peculiar to living matter,
or bioplasm.

4.) Bioplasm can also transmit vital power to its prog-

eny.   This property will be considered more in detail in
the next chapter, on Parentage.

8. The peculiar relations and changes of the chemical
elements in bioplasm prove it to possess some power
different from not-living matter, whose actions or results
no chemistry can predict.   We have said that bioplasm
consists chemically of oxygen, hydrogen, carbon, and
nitrogen.   Other unessential elements may also be present in some cases.   But we cannot tell how these elements are combined, if, indeed, they are combined at all
in the proper sense of that word.   As all bioplasm presents the same appearance, although differently formed
material results from its transformation — different in
physical properties and in chemical composition — as
muscle, nerve, bone, etc., it is probable that the elements do not combine at all as in inorganic matter, but
that the ordinary affinities are suspended or modified
by vitality.

Bioplasm is a semi-fluid substance, yet it will not
freeze at 32° F., as water does, showing that in this respect it is different from water.

Bioplasm is in a state of constant molecular change,
or unstable equilibrium, since it is constantly receiving
pabulum and transforming itself into formed material,
so that it is doubtful if chemical combination is possible
during life, the atomic activities being too transitory for
combination.

When change takes place from bioplasm into formed material combination occurs, but the formed material is not
living tissue, or bioplasm.   The life is gone.   It is dead,
as if it had never formed part of an organism, although

it may have acquired special properties, as the elasticity of muscle, or the conducting power of nerve tissue.

If the change referred to occurs suddenly, that is, if the life of bioplasm is suddenly destroyed, the result is water, albumen, fat, and sometimes fibrin, and certain salts, as chloride of sodium, etc.

In slower transformations, which are equivalent to slow molecular death, different materials result, as fat, sugar, milk, biliary acids, etc. Free oxygen is sometimes absorbed, and very complex compounds result, often baffling analysis.

Physiological Chemistry has traced many of the results of changes in formed material, but the composition and physical surroundings of germinal or living matter will not indicate the nature of its transformations nor its function. No one can tell whether a particular bioplast belongs to a vegetable or an animal, whether it will form an eye or a finger, a nerve or a piece of bone, nor whether its function shall be secretive, excretive, elastic, or conductive. Nothing but observation can tell its future life-history.

9. Although all bioplasm has powers or endowments which transcend all physics and chemistry, and which can only be accounted for by that dualistic philosophy which acknowledges the reality of both matter and spirit, yet "all flesh is not the same flesh." There is an original and essential distinction between bioplasts. The bioplasm of a fungus never produces a fish, nor that of a butterfly a man. This will be fully discussed in the chapter on Parentage. Yet it is no easy task to discriminate between living forms, especially in what are

called the lower orders. It is difficult to distinguish in all cases between animals and plants. In the simpler kinds the characters touch and dissolve into each other, so that no exclusive definition is possible. Some naturalists think that there are organisms which at one period of life are vegetable, and at another animal.

10. If we consider their origin, both animals and plants begin life as a small particle of bioplasm. In plants this forms an ovule, with wall of cellulose, and in animals it becomes an ovum, or egg, with wall of albuminous matter.

11. As to form, we have no means of separating animals and plants. The zoospores of Algæ are like Infusoria. Sea-mat (Flustra) and Sea-moss (Fig. 4) (Poly-zoa) are like Sea-weeds, (Algæ,) Corals and Actiniæ are like flowers.

FIG. 4.—Sertularia Operculata.

12. In chemical composition, as a rule, plants are destitute of, and animals are largely supplied with nitrogen. Yet there are some animal structures without nitrogen, and some vegetable structures with it. Cellulose, (woody fiber,) generally found in vegetables, is wanting in the Fungi, and is found in the covering of Ascidians, (Sea-squirts.) Starch, under the name of

Glycogen, is found in the liver and in the brain. Chlorophyll, which makes the leaves of vegetables green, is found among animals, as in Stentor, (the trumpet-shaped animalcule,) and in Hydra viridis, (the green hydra.)

13. As to locomotive power, bioplasm is essentially active, as I have described, both in plants and animals. The zoospores of Algæ are covered with cilia, and move in water like animalcules. Motion is common among Diatoms, Desmids, Oscillatoria, and other classes of plants, while Sponges, Corals, Oysters, and Barnacles are largely destitute of locomotive power.

14. With respect to food, plants live generally on mineral or inorganic matter, chiefly water, carbonic acid, and ammonia, while animals require ready-made organic compounds to support life. Thus plants manufacture and animals consume organic pabulum. Yet Fungi, which are generally classed with vegetables, feed as animals on organic matters, and insectivorous plants, as Darwin has shown, feed on animals.

15. Animals generally possess sensation, consciousness, and volition, yet there is a kind of sensation in the sensitive plant, Venus' fly-trap, etc., and something like volition in zoospores, or they would often collide in the active dance they keep up. Plants need rest as well as animals. Both have their epidemics, poisons, and remedies.

16. If we admit a dualism, or spiritual cause of life, in vegetables, as well as in animals, it does not prove them immortal. Immateriality does not imply immortality. Existence, spiritual or material, depends on the will of

the Creator, and we can only know the future as he has
revealed it.

> "Heaven from all creatures hides the book of Fate,
> All save the page revealed—the present state."

17. Our study thus far impresses us not only with the
truth that all living things manifest a dualism, but also
that all living are intimately related. Not that all come
from a single germ, or from a few germs, but that animals and plants form, after some sort, a common family.
From the great Father and Fountain of life all living
things proceed, and their existence and endowments are
according to his will. Immaterial, or spiritual existences weave for themselves a beautiful garment from the
inorganic world. The plant bioplasm appropriates mineral matter, with carbonic acid, water, and ammonia,
and by a wonderful vital chemistry transforms it into
organic compounds, as starch, sugar, gum, albumen, etc.
These compounds afford pabulum to animal bioplasm,
and are transformed to blood, muscle, nerve, and other
complex animal substances. After these transformed
products have served the purposes of animal life they
are discarded, and return again to the mineral world.
Thus the wonderful wheel of life revolves from age to
age under the watchful care of divine Providence.

18. The intimate relations of living things may find a
mathematical illustration in the logarithmic spiral, such
as is described by a ship sailing N. E. at an angle of 60°
from the pole. It is the *spira mirabilis* of Jas. Bernou-
illi, who desired one to be engraved on his tomb, with
the motto: "*Eadem mutata resurgo*"—"I rise the same,
though changed." It is a spiral which has the same

4

character in all its parts, and which may continually decrease in the size of its windings without coming to a point, or increase the number of its convolutions to infinity. Such a spiral may illustrate the continuity, yet varying amplitude, of creation. We may trace the progressive windings of creative power from the motions of inorganic bodies in space to the motions of bioplasm in the vegetable world and to the higher nerve-structures of animal life. In all organic matter we see the workmanship of the same Great Artist:

> " Lo ! on each seed within its slender rind
> Life's golden threads in endless circles wind ;
> Maze within maze the lucid webs are rolled,
> And, as they burst, the living flame unfold."

In exact truth, however, each widening circle of creation exhibits some new and higher form of creative power and skill. The circle widens, and is also in another plane. Something has pushed forward the center. Every spiral requires a progressive force, as well as a centripetal and centrifugal one. Each specialization—either elevation of type or specific difference—involves new force-expenditure. Certain factors have been successively added. First, we find inorganic matter, of many kinds, or of a single kind. Next, the physical forces, so-called, but really the activity of a personal Creator on the matter he has formed. Then we find life, or the activities in matter of created spirits in most wonderful gradation. Rising to another plane we find added to this life mind-force, or intelligence. Still higher we find spirit, properly so-called, possessed with moral properties, giving dignity to men and angels.

Yet the spiral is not broken, it is but expanded, and the analogies and relations have a distinctive similarity, since they are equally the work of one God and Creator of all. As the physical forces, by attraction and vibration, and conservation, arrange the cosmos, or physical universe, so the various bioplasts weave the living tissues for the living creature—the microcosmos—and so the conscious acts of our spirits weave the character of our future life.

CHAPTER III.

PARENTAGE.

We must get rid of all these complications of an erring philosophy, this floating chaos of mist and phantasms, and return to the Natural Realism, which all men have been learning from their first hours of childhood, and can never unlearn, before a science of Physics can be really founded. Its first principle is that we are real persons, living amid a real world of material objects distinct from ourselves. And this double truth leads upward to One who is the cause both of matter and mind, the Supreme Reality, who dwells in light inaccessible, but who can reveal himself, and has revealed himself, in love and mercy to the souls he has made. — *Modern Physical Fatalism*, by T. R. BIRKS.

1. TWO theories divide the learned world respecting the genesis of living things; the doctrine of parentage, or the descent from living creatures each created "after his kind," and the theory of spontaneous generation of the living from the non-living, and the transmutation of one kind of living beings into another. The first theory is sometimes called the doctrine of Creation, the latter that of Evolution.

2. The word Evolution simply means *to unfold*, and may be used to express the life-history of individuals or of species, or the development of the plans of the Creator in the natural world. To such a meaning there would be no objection by any one, but as it is generally understood to mean the mechanical or monistic view of the universe, which ignores a Creator, and teaches the eternity of substance, the invariability of law, and the transmutation of living beings, its use should be re-

stricted to that view.   Any other application of it leads
to confusion of thought.

3. There is nothing new in the modern doctrine of
Evolution.   Among the Greeks, Leucippus, Democritus,
and Epicurus taught that all forms and phenomena came
from the spontaneous motions of atoms, and this view,
in all probability, was a product of older Indian pan-
theism.

Modern upholders of transmutation differ from each
other greatly in the details of the theory.   Some are
atheistic, or agnostic, leaving the Creator entirely out of
view.   Among these, some teach, like Lamark, the self-
elevation of species by appetency, or desire, use, and
effort.   Others, as Darwin, Haeckel, and many late writers, teach what is called natural selection with spontaneous variability, or the survival of the fittest.   Others
again, as Draper and Spencer, teach modification of
species by the surrounding conditions.   Some evolutionists are deistic, like Owen and Mivart, and teach a pre-
ordained succession, under internal force or innate
tendency; or, as Morell and Murphy argue, evolution
by unconscious intelligence.   In opposition to these
views the majority of naturalists of this and the past age
hold to the doctrine of parentage, and deny the change
or transmutation of species, although admitting a certain amount of physical variability, producing races or
varieties.   Among these may be named Linnæus, Cuvier,
Agassiz, Dana, Guyot, M'Cosh, Balfour, Dawson, Milne,
Edwards, and Seelye.

4. The acknowledged ability of Agassiz in regard to
all matters connected with natural science entitle his
'4*

opinions to careful consideration. He says: "It is my
opinion that naturalists are chasing a phantom, in their
search after some material gradation among created beings, by which the whole Animal Kingdom may have
been derived by successive development from a single
germ, or from a few germs. . . . It is contradicted by the
facts of Embryology and Palæontology, the former showing us norms of development as distinct and persistent
for each group as are the fossil types of each period revealed to us by the latter." "If they are linked together
as a connected series, then the lowest Acaleph should
stand next in structure above the highest Polyp; and
the lowest Echinoderm next above the highest Acaleph.
So far from this being the case, there are, on the contrary, many Acalephs which, in their specialization, are
unquestionably lower in the scale of life than some
Polyps, while there are some Echinoderms lower in the
same sense than many Acalephs." He shows that the
same principle applies to classes in other types: "There
are some members of the higher classes that are inferior
in organization to some members of the lower classes."
The same thing is true in Embryology: "A Vertebrate
never resembles at any stage of its growth any thing but
a Vertebrate, or an Articulate any thing but an Articulate, or a Mollusk any thing but a Mollusk, or a Radiate
any thing but a Radiate." Geologically, also, we see no
transition between types. "In the earliest fossiliferous
strata there were the three classes of Radiates, two of
the classes of Articulates, and one of the classes of Vertebrates." The Geographical Distribution of animals
proves the same thing. Thus Agassiz proves that the

Series of Rank, of Growth, of Time, and of Geographical Distribution all show that there is no such gradation
as transmutation implies, and that the connection between different kinds of living things is not a material
connection, but only an intellectual one, indicating the
plan of the Great Architect.*

5. In all forms of life which have yet come under
human observation, the origin has not been by transmutation, but by parental derivation. Animals and vegetables all come from parents of similar organization. If
ever transmutation was the law of origin, it has been
changed, and the law of parentage is now supreme. But
a change of law is inconsistent with the theory of evolution. Unless the law had been changed, species would
still originate by transmutation, if ever they had such
origin. Such transmutation has never been observed.
The Egyptian monuments prove that the animals of
earliest history remain unchanged, and Agassiz has
shown from the coral reefs in Florida that the animals
of the Gulf of Mexico remain the same as when these
deposits began. Even the varieties which man secures
by "artificial selection" revert to the original type when
the modifying circumstances are removed. Transmutation has not a single fact to prove it. At best it is but
a theory, and one which all the facts known render most
improbable.

6. The geological evidence shows the entire absence
of intermediate varieties between species, which intermediate forms Mr. Darwin himself admits to be necessary to establish his theory of natural selection. He

* Agassiz, " Methods of Study in Natural History"

claims that the geologic record is defective, and that
when it is better known it will exhibit these forms. But
among more than 30,000 species, many of them represented by thousands of individuals, some of the intermediate forms would occur, if any ever existed. Professor
Pfaff has shown the improbability of the terminal links
only of the chain being preserved by applying the calcu-
lus of probabilities. If 100 individuals of each species
have been found, and 10 intermediate varieties existed,
(a smaller number than Darwin claims,) the probability
against the exclusive appearance of distinct species is as
$1:10^{100}$, ($1:1$ with 100 ciphers annexed.*) Professor
Marsh claims to have discovered apparently intermediate forms between the Palæotherium and the horse, but
the proof that the Palæotherium, or the bones referred
to, belonged to the progenitors of the horse has not
been shown, any more than the juxtaposition of bones
of the horse, the zebra, and the ass, would prove them
to be derived from each other. If it were proven, although it would show great variability in that species, it
would not establish transmutation.

7. Geology shows that some of the first forms of life
are also the latest, as the corals. If transmutation be
true, in the struggle for existence they should have
disappeared by being changed into something higher.
That they have not makes against Evolution.

8. Believers in transmutation claim that all living came
into existence by the gradual modification of a primitive
germ, and they find plausibility for this in the development of a single bioplast into the various tissues of an

* Johnson's "Cyclopedia, Art. Darwinism."

animal. Another analogy is found in the development
of the embryo. As the tadpole is first a fish, and then
a tailed amphibian with lungs and gills, before it becomes a frog, so they deem that the history of the
embryo recapitulates the transformations of the species.
This sort of theorizing has given rise to numerous efforts
to arrange the family tree of each species—a branch of
biological speculation termed *Phylogeny*—and examples
of it may be found in Darwin, Haeckel, etc. Mr. Huxley, although a believer in Evolution, declares that such
summaries of descent are little better than guess-work.*

9. Many instances of complicate and perfect structure
occur both in the vegetable and animal kingdoms which
have no similar structure preceding nor following them.
No scheme of evolution, nor survival of the fittest, can
account for them. The mechanism of the leaf of Venus's
fly-trap, and of Nepenthes, the nettling threads of Hydroid polyps, and the peculiar disk-like hairs on the
thigh of the male water-beetle, (*Dytiscus marginalis*,) are
a few out of almost numberless instances of this fact.
The most perfect dental apparatus in the animal kingdom, the teeth of Echinus, called Aristotle's lantern, is
also the first to appear, if we trace animal life from its
simplest forms, and there is nothing like it elsewhere.
Like Melchizedek among priests, it has no predecessor
and no successor. Its form and arrangement are a pro-
test against the theories of material development. In
the Rotifer, again, the typical form and structure of the
teeth are entirely different, being an anvil and two hammers. In the Gasteropods they are spiny tongues.

* "Anatomy of Invertebrated Animals."

10. Evolutionists find it difficult, if not impossible, to account for the first origin of living matter. The boldest and most logical among them maintain that it began spontaneously out of non-living matter. Some, like Sir W. Thompson, suppose that the germs of living things were transported to our globe from some other. Others, as Darwin, hold to the creation of a single germ, or a few germs, from which all have developed. The doctrine of the spiritual origin of living things is beset with no such difficulties as the mechanical theory. While it admits a unity of plan resulting from the superintending intelligence of an all-wise Creator, it sees in living things a true diversity also. It is hard to imagine how a naturalist can think of "differentiation" without acknowledging a cause of variety *ab extra*, (from without.)

11. The evidence adduced in favor of spontaneous generation is always of one kind. A quantity of animal or vegetable infusion is boiled in a flask, which is then hermetically sealed. After a time minute forms of life are found on a microscopic examination of the fluid. It is taken for granted that all living germs are destroyed by boiling water, and that therefore the organisms seen after a few days are developed spontaneously. But Messrs. Döllinger and Drysdale have shown that some germs remain alive after exposure to a temperature of 300° F., and Pasteur has found that stopping the necks of the flasks with cotton wool, so as to filter the air from all germs, prevents the appearance of Infusoria, as well as of decay, in fluids well adapted to such organisms. Professor Tyndall has also experimented with a great variety of fluids in air so deprived of floating germs

as to be optically pure, and has had similar results. So that we may consider the question to be scientifically settled, and that all living beings come from similar parentage, or, as Virchow expresses it, "*omnis cellula e cellula*," (every cell is from a cell.)

12. Parentage is of two kinds, sexual and non-sexual. In the first, we sometimes find the sexes distinct, as in the higher animals, and sometimes united in the same individual, as in the stamens and pistils of most flowers, and as in some animal forms.

Non-sexual generation is seen mostly in the simpler forms of animal and vegetable life, and as it throws light on many of the phenomena of nature which would otherwise be obscure, we notice this form of reproduction here in a general way, reserving special instances until we treat of the life-history of each class.

13. In referring to the Vorticella, or bell-shaped animalcule, in our first chapter, mention was made of its increase by self-division. The mass of bioplasm of which it is composed separates into two masses, which become separate individuals. This mode of increase is called *Fission*, and is quite common among the minuter forms of life. In *Sarcina ventriculi*, a sort of vegetable parasite, the division is into fours, or four times four.

14. A variety of fission, called *Gemmation*, or *Budding*, is often met with. A portion projects from the mass, and separates to begin an individual existence. Thus in the fresh-water polyp, or Hydra, a bud gives rise to an organism like the parent, which becomes detached and independent. Sometimes the product of buds remains attached, as in plants, and in the Foram-

inifera. In other cases the budding is internal, and the progeny may or may not remain attached to the parent.

15. *Alternation of generations* is a term given to express a mode of reproduction in which "the parent finds no resemblance in his progeny until he comes down to his great-grandson." The Jelly-fish, or Medusæ, from the huge masses cast up by the waves of the sea-shore, to the tiny bell no bigger than a pea, are developed in this manner. A ciliated germ, like some of the Infusoria in form, swims about awhile, then becomes attached, elongates, and develops into a polyp like the Hydra. The polyp becomes wrinkled and subdivides until it looks like a pile of saucers with scalloped edges. This breaks into segments, each of which becomes a jelly-fish, which enlarges and produces fresh germs. Fig. 5. This form of reproduction differs from metamorphosis, such as a butterfly

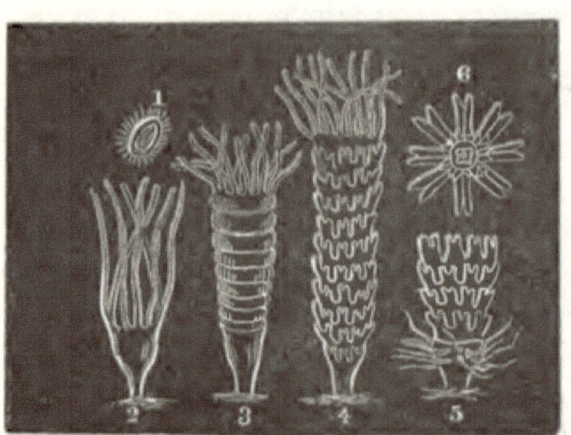

FIG. 5.—Diagram illustrative of the Development of Hydrozoa.
(The specimen is one of the Lucernaridæ.)

1. Ciliated embryo or "planula." 2. Hydra tuba, showing a single individual. 3. Hydra tuba undergoing segmentation. 4. The segmentation becoming more complete. 5. More advanced stage, in which the tentacles are developed from the first or basal segment. 6. Segmentation complete, giving rise to a free swimming Medusoid.

undergoes in passing from the egg to the perfect insect, or as most animals pass through in the embryonic state.

The caterpillar becomes a butterfly, but the hydra-like individual referred to produces a number of Medusæ.

16. *Partheno-genesis*, or virgin production, denotes the production of new individuals by virgin females without the intervention of a male.

The Aphides, or plant lice, so often found parasitic on plants at the close of autumn, consist of winged males and wingless females. The ova, or eggs, are dormant through the winter, and the young hatched in the spring are sexless, but produce viviparously a brood like themselves, and this generation produces another, and so on for ten or twelve generations, the last brood being male and female as at first. Many other tribes of insects afford examples of partheno-genesis.

17. The subject of this chapter brings us to some of the deepest mysteries of creation. The parentage of all living, and the various modes in which the principle of parentage is manifested—such topics are wonderful seed-thoughts. It is not likely that we shall ever understand fully the repetition of individuality, but we see enough to indicate some of the plans of the Designer of all. "Lo! these are parts of his ways . . . but the thunder of his power who can understand?"

Some analogies between the teachings of biology as to the genesis of living things, and some of the statements of Scripture, may be readily traced. Mr. Joseph Cook has been sharply criticised for comparing the birth of Jesus, as revealed in the Gospels, with partheno-genesis; yet he had reason for so doing, nor is he alone in his opinion. In President Dawson's "Origin of the World" we read, "It is curious that the Bible suggests three

methods in which new organisms may be, and, according
to it, have been, introduced by the Creator.   The first is
that of immediate and direct creation, as when God created the great *Tanninim*, (whales.)   The second is that
of mediate creation, through the materials previously
existing, as when he said, 'Let the land bring forth
plants,' or 'Let the waters bring forth animals.'   The
third is that of production from a previous organism by
power other than that of ordinary reproduction, as in
the origination of Eve from Adam, and the miraculous
conception of Jesus."—P. 229.

"The Bible indicates some ways in which living creatures may be modified, or changed into new species, or
may give rise to new forms of life.   The human body is,
we are told, capable of transformation into a new or spiritual body, different in many important respects, and the
future general prevalence of this change is an article of
religious faith.   The Bible represents the woman as produced from the man by a species of fission, not known
to us as a natural possibility, except in some of the lower
forms of life.   The birth of the Saviour is represented as
having been by partheno-genesis, and if it had pleased
God that Jesus was to remain on earth as the progenitor
of a new and higher type of man to replace that now existing, this might be regarded as the introduction of a
new species."—P. 378.

It certainly disarms skepticism and strengthens the
probability of Bible history, to find such analogies between the natural world and the record of revelation.

Living beings are not fortuitous nor necessary group-
ings of atoms, either mechanical, as Monism teaches, or

monads of force, as Leibnitz wrote, but sparks of spiritual existence, given off voluntarily from the Eternal Parent, having various powers and capacities, yet each capable of pressing the fleeting atoms of matter into its service during the period alloted to it in the world. Of all living beings man is nearest like the Great Father, in whose image we were created, and who, when heart and flesh—body and animal life—shall fail, may be the strength of our hearts and our portion forever.

" For we also are his offspring."

CHAPTER IV.

TISSUE FORMATION.

In regard to the physical universe, it might be better to substitute for the phrase " government by laws," " government *according to* laws," meaning thereby the direct exertion of the Divine Will, or operation of the First Cause in the Forces of Nature, according to certain uniformities which are simply unchangeable, because, having been originally the expression of Infinite Wisdom, any change would be for the worse.—DR. W. B. CARPENTER.

1. A TISSUE is a structure which presents a special form and serves a special purpose. Thus we find in plants cellular and woody tissues, and in animals muscular, nervous, connective, and epithelial tissues, etc. From tissues are formed organs, as the circulatory, respiratory, or digestive organs. A collection of organs serving a common purpose is called a system, as the nutritive, generative, or nervous systems. The union of systems in a co-ordinate organism, or the equivalent of such a union, forms an individual. An individual among the higher forms of life is a very complex arrangement of systems and organs; but in the lower forms more simple arrangements prevail, which may be considered equivalent, or representative, of complicated organs, as in the Rhizopods, referred to in Chap. II., Sec. 5.

2. In the formation of tissues, the peculiar living properties of bioplasm already described ; the physical agencies of light, heat, electricity, and moisture ; chemical reactions such as are common to inanimate substances ;

and certain properties called *osmose* and *molecular coalescence*, all combine, so as to render the study of some tissues quite complex. In other cases the mode of formation is readily traced.

3. The action of physical stimuli, as heat, etc., upon bioplasm itself is yet very imperfectly known. Light is not essential to its development, as is seen in the growth of fungi, the cells of the interior of organisms, and of the embryo in the dark. Many experiments on bioplasm have shown that a moderate increase of temperature quickens its movements, and a corresponding depression retards them. Electrical, mechanical, and chemical stimulation have similar effects to heat. Yet the action of these stimuli vary in different cases. The motions of amœbæ are arrested by iced water, and recommence on raising the temperature, yet the segmentation of trouts' eggs proceeds well in iced water, but in a warm room they soon die.* If the change of intensity in the stimulation be made gradually, and not suddenly, the living matter will sometimes adapt itself to it without serious disturbance. Animals have been frozen and revived, and there are instances on record of men enduring for a considerable time without much inconvenience the heat of ovens raised to 500° F.

The influence of light, heat, and electricity upon formed material of different kinds is very great, but the complexity of the organism and of the phenomena render it difficult to know what part is supplied by the bioplasm and what by its product. The vegetable bioplasm of the interior grows and reproduces its kind, but the

* Stricker's " Manual of Histology."

5*

green chlorophyll which it forms beneath the epidermis, especially in the leaves, under the influence of light alone breaks up carbonic acid for the supply of carbonaceous food. The influence of the more luminous rays, as the yellow and orange, is greater in this respect than the others. Gardeners blanch certain plants by raising them in the dark, yet in the first part of the germination of seeds Light is injurious rather than beneficial. The influence of Light upon the direction of the growing parts of plants, the opening and closing of flowers, etc., may be chiefly owing to its influence upon the chlorophyll referred to above, or it may be in some degree a direct mechanical stimulus. The same amount of Light, however, is not required for all plants. Some require a very different amount than others. Among animals Light has considerable influence upon colors, and still more upon the process of development. Persons who live in cellars or in dark streets are apt to produce deformed children, while recoveries from disease are promoted by the access of light.

To every species of plant and animal there is a congenial and favorable temperature, although great varieties exist in this respect, as well as in the power of adaptation to extreme conditions. Many plants, for example, perish with the slightest frost, yet the little fungus (Torula) which is the principal agent in yeast, does not lose its vitality at 76° below zero, although requiring a somewhat elevated temperature for its active growth.

Electricity possesses the power of exciting the contractility of the muscular fibers and the nervous force in animals in a remarkable degree. It has, however, mechanical, chemical, and thermal influence, in addition to its own special power, so as to be a very valuable agent in scientific medicine; yet the nature of its relation to the living organism is not yet understood.

In every organized being there is an incessant play of most varied actions. Buffon well said, " The animal combines all the forces of nature; his individuality is a center to which every thing is referred, a point reflecting the whole universe, a world in miniature." It is a one-sided philosophy, however, which sees in the living thing nothing more than the forces which are outside of it and play upon it, and are, to a great degree, subject to it.

4. *Osmose*, or osmotic action, is a property of animal and vegetable membrane, and of some other porous or soft materials, by which liquid substances may be separated from each other. If two liquids (or gases) capable of mixing with each other are separated by paper, caoutchouc, or a bladder, one liquid being suspended in a bladder, or in a cylinder with its lower end tied over with bladder, etc., and immersed in the other liquid, the liquid within will pass through the bladder into the other, (*exosmose*,) or the liquid without will pass into the bladder, (*endosmose*,) or both endosmose and exosmose will take place at the same time until there is an equal proportion of liquids on either side. (Fig. 6.) These phenomena are owing to the physical attraction

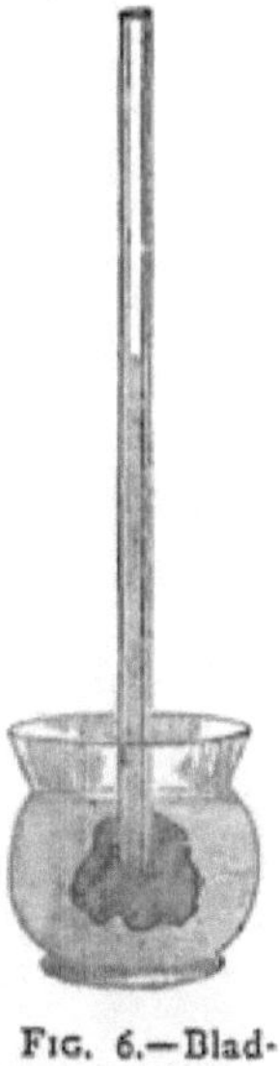

FIG. 6.—Bladder containing syrup, attached to a tube and plunged in a vessel of water. The inward motion of the water (endosmose) exceeds the outward movement of the syrup, (exosmose,) and presses the fluid up the tube.

the two liquids have for each other and for the membrane separating them.

Crystallizable bodies, as salt, niter, etc., when in solution, and substances allied to them, as hydrochloric acid, and alcohol, pass readily through membrane; but bodies which do not crystallize, but assume the gelatinous form, as gum, starch, albumen, hydrate of alumina, etc., pass through, if at all, with great slowness. Such bodies are called *colloid*, or glue-like. Osmose occurs through all jelly-like bodies, as bioplasm, as well as through fully formed membrane, and in this manner various liquids are absorbed or imbibed by the tissues.

5. *Molecular coalescence* is a term applied to the modification of ordinary forms of inorganic particles which occurs when they combine in the presence of organic matter. Thus it has been found that the crystallization of certain salts of lime, as the carbonate, when occurring in a solution of some organic colloid, as gum-arabic, albumen of eggs, blood-serum, and gelatine, is so modified by such a solution as to resemble many of the calcareous deposits found in nature.

The bottom of the middle and northern parts of the Atlantic Ocean is found by the deep-sea dredge, even at the depth of nearly three miles, to be covered with a sort of slimy ooze, which Prof. Huxley formerly deemed to be of animal nature, and termed *Bathybius*. More recent investigations have led him to change this opinion. It is regarded as a gelatinous inorganic secretion, or a product of Diatoms, a family of minute Algæ. In this slime great numbers of globular, shell-like microscopic masses are found, similar to those in the chalk strata of the earth's crust. By experiments in molecular coalescence similar forms have been produced artificially.

Spicules, like those in the skin of certain marine animals, have also been formed by molecular coalescence, as well as laminated plates like cuttle-fish bone. It is quite probable that many calcareous deposits in tissues, as in the shell of the bird's egg, in the scales of fishes, as well as in bone and teeth, may be thus accounted for. The presence and contact of living colloid matter modifies the ordinary laws of crystallization, and produces forms differing according to the endowment of the bioplasm.

6. In vegetables most of the organs are composed of *cellular tissue*, or a congeries of cells. The surface of the cell, which originates by fission from bioplasm, is changed into *membrane*, or cell-wall, while a *nucleus*, (one or more,) now generally regarded as a concentration of vital power, appears inside. Within the nucleus, another spot, the *nucleolus*, is sometimes seen. (Fig. 7.) The cell itself presents the appearance of a bladder full of

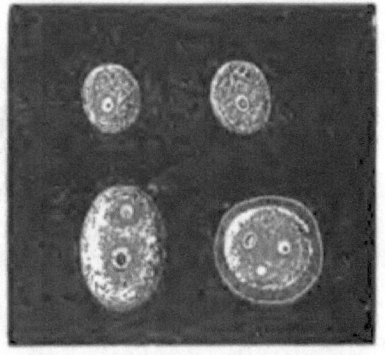

FIG. 7.—Vegetable cell, with nucleus and nucleolus.

fluid or semi-fluid material, in the midst of which the nucleus is visible.

7. Many simple vegetable forms consist of a single cell, the membranous wall of which is a species of formed material called *cellulose*, a substance analogous to starch.

Within this membrane the bioplasm is, as it were, imprisoned, yet receiving pabulum by endosmose, or through pores left in the membrane, its vital functions remain. In the higher plants, as the palm or the oak, the structure is but an aggregation of cells, some of which have been modified in form to serve special uses.

8. Near the vegetable cell-wall the bioplasm appears less fluid than in the middle of the cell, and certain chemical agents cause a partial separation from the membrane, so as to present, under the microscope, the appearance of a secondary and gelatinous membrane—the *primordial utricle.*

In some vegetable cells the molecular movement of the contained bioplasm is quite evident, and has received the name of *Cyclosis.* It may be seen under the

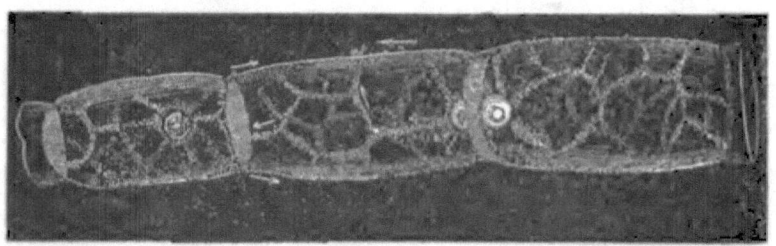

FIG. 8.—Three cells from the hair of a potato, showing Cyclosis. Bioplasmic threads proceed from the nuclei, along which the current flows, in the direction of the arrows.

microscope in the stinging hair of the nettle, and in hairs from the calyx of *Tradescantia Virginica*, etc. (Fig. 8.)

9. Within the cell-wall the bioplasm may be transformed into chlorophyll, or green coloring matter, into starch, gum, oil, resin, sugar, or other kind of formed material or mineral substances may crystallize in the cells, forming what are known as raphides. The variety

of vegetable products of this kind is very great. (Figs. 9 and 10.)

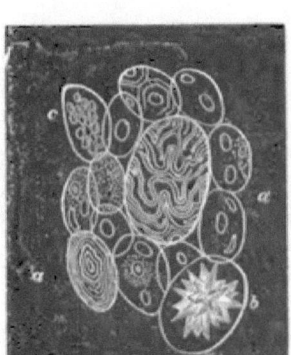

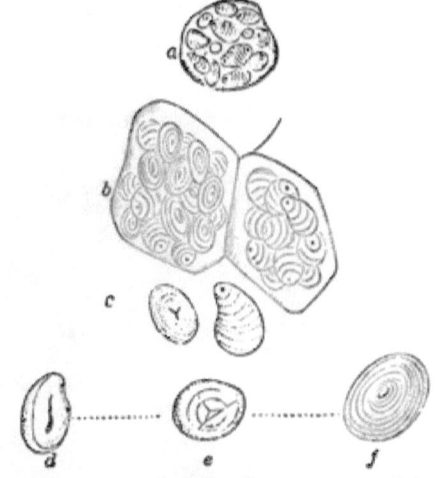

FIG. 9.—Cellular tissue of Cereus variabilis, containing: *a. a.* Jelly. *b.* Crystals. *c.* Starchgranules.

FIG. 10.—*a. b.* Cells of a potato, containing starch. *c.* Starch-grains apart. *d. e. f.* Wheatstarch in different positions.

10. There is often a deposit of silica on the cell-wall, as in grasses, horsetails, and diatoms. Some of these latter are beautifully marked with lines and dots, rivaling the most complicate patterns of engine-turned engraving.

11. Cell-membrane, as all other kinds of formed material, grows by addition inside, so that the inner layer is the youngest. The formed material may get so thick that nutrition ceases and the bioplasm is wholly transformed, or dies. The solid deposit which fills up the cells of woody fiber is known as *sclerogen*, or woody tissue. (Fig. 11.) In Coniferous plants the fibers are

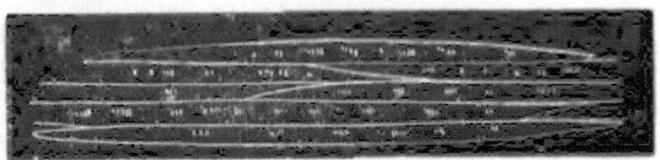

FIG. 11.—Woody fiber.

marked with depressions, or concave spaces, (glands?) the centers of which are penetrated, as if some sort of special communication existed between the bioplasm

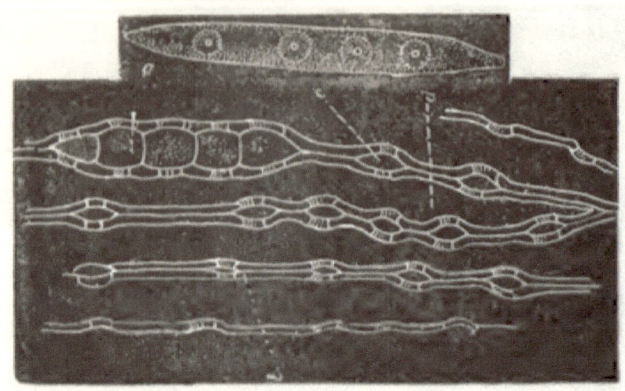

FIG. 12.—Glandular fiber. *a.* External appearance. *b.* The sides of two tubes, or fibers, in contact. *c. d.* Lenticular cavity between the tubes.

of contiguous cells. (Fig. 12.) Sometimes sclerogen is deposited within the cell-wall in such a manner as to

FIG. 13.—Annular and dotted cells.

produce dots, or pores, or rings, or spiral fibers, which give names to the several kinds of cells. (Fig. 13.)

12. Vegetable cells are of various shapes, according to the purposes they subserve. They may be conical, oval, prismatic, cylindrical, sinuous, branched, entangled, or stellate. (Fig. 14.) Tubes, or vessels, are formed of elongated cells. Sometimes such cells join end to end, and the partition being removed by absorption, a long tube is formed. Such vessels may be dotted, reticulated, annular, or spiral, from the deposit of woody tissue, or sclerogen. (Fig. 15.) In the stem of Endogenous plants, as palms, etc., bundles of fibro-vascular

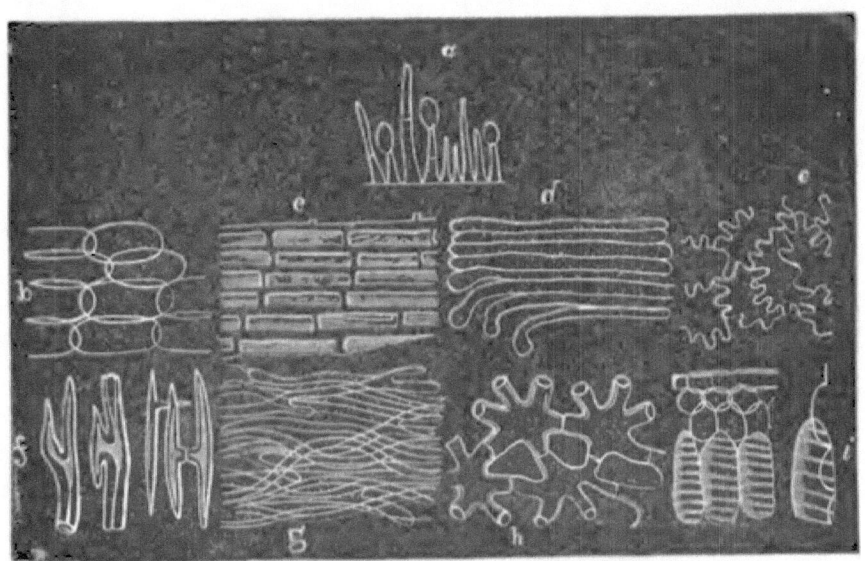

FIG. 14.—Various forms of cells: *a.* Conical. *b.* Oval. *c.* Prismatic. *d.* Cylindric. *e.* Sinuous. *f.* Branched. *g.* Entangled. *h.* Stellate. *i.* Fibro-cellular tissue.

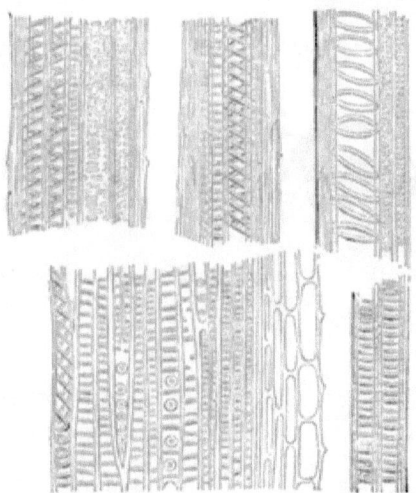

FIG. 15.—Annular, dotted, and spiral vessels and ducts.

tissue occur among a mass of cellular tissue; but in Exogens, as the maple, oak, etc., we find a more regular arrangement of pith, medullary sheath, wood, bark,

6

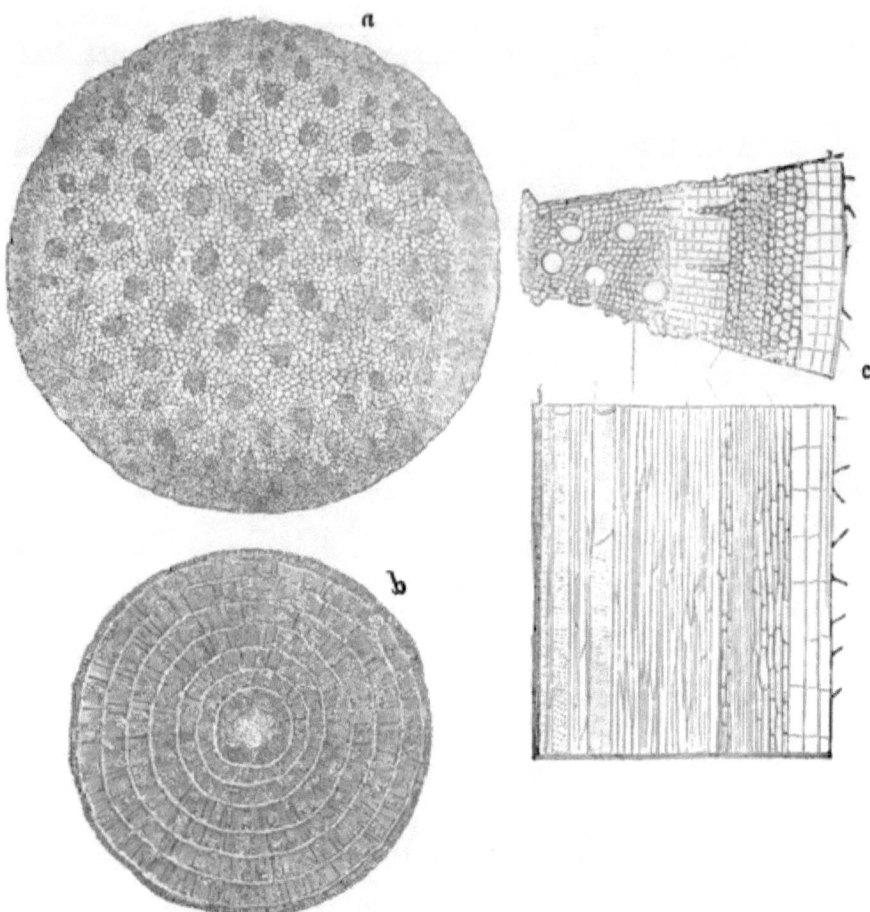

Fig. 16.—*a*. Transverse stem of Endogen, (Palm.) *b*. Of Exogen, (Buckthorn.) *c*. Transverse and longitudinal section of Maple in the beginning of the second year

and medullary rays. (Fig. 16.) The pith is the cellular tissue of the center; the medullary sheath a ring of spiral vessels round the pith, which sends projections through it to form the medullary rays; the wood consists of concentric layers of woody and vascular tissue; and the bark is made of cellular materials, sometimes containing branching vessels (laticiferous tissue) conveying milky juice.

13. Leaf-tissue is made up of cells, with cavities, fibro-vascular bundles, and epidermis. (Fig. 17.) The

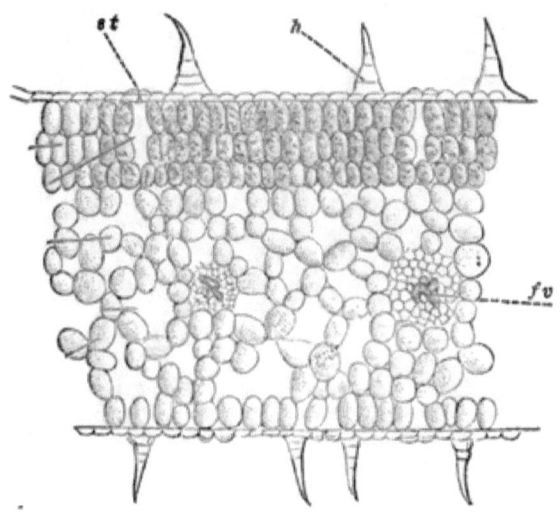

Fig. 17.—Perpendicular section of Melon-leaf: *h.* hairs; *st.* stomata; *fv.* fibro-vascular tissue of the veins.

latter is a sort of skin composed of compressed cells, among which are found openings, or pores, (*stomata,*)

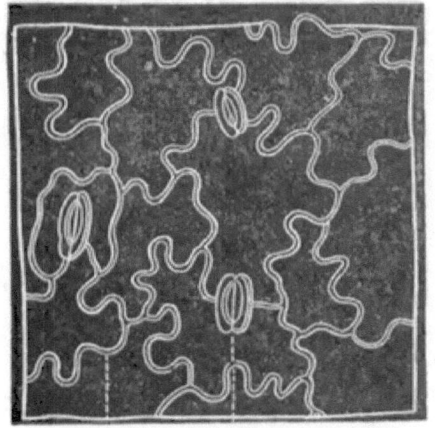

Fig. 18.—Epidermis of Madder, with stomata.

each guarded by two or more elastic cells which regulate evaporation and respiration by their expansion. (Fig. 18.)

From the surface of the epidermis arise hairs, formed of minute expansions of cellular tissue. They are of various forms. Some of them secrete volatile oil, others, as the nettle, an acrid fluid. They often form microscopic objects of great beauty. (Fig. 19.)

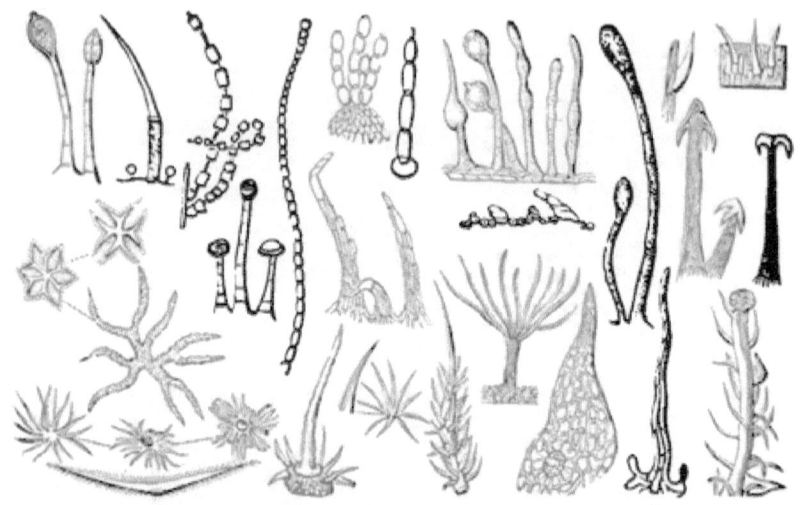

FIG. 19.—Various forms of vegetable hairs.

The poet Goethe first clearly showed that the various parts of the plant, from the seed to the blossom, are but modifications of the leaf. All the parts of a flower, calyx, corolla, stamens, and pistil, are only leaves adapted for peculiar functions. They were not originally leaves, and afterward transformed, but they are formed of the same elements, and arranged upon the same *plan*, and in the changes which they undergo and the relation they bear to each other, they follow the same laws as leaves do.

All leaves are arranged upon the stem after two leading patterns—the *whorl* and the *spiral;* but as by teasing out the whorl we get the spiral, and by compressing the spiral we get the whorl, we may regard them as essentially the same.

14. In the animal kingdom, with the exception of

those simple forms of life already described, which increase by fission or budding, (Chap. III., Sec. 12, 13,) the
germ of all the tissues is first a piece of simple bioplasm
derived from the vesicles of the ovary. This is fertilized
by fusion with similar bioplasm derived from the male.
It then acquires a membrane, and exhibits a nucleus
and nucleolus, as in the case of the primitive vegetable
cell. Changes, however, take place in the animal ovum
which we do not observe in the vegetable, and these
changes differ also in the different classes of animals.
In the higher classes the ovum separates into two
spheres, which sub-divide into four, then into a mul-
berry-like mass of cells, or *morula*. (Fig. 20.) These cells

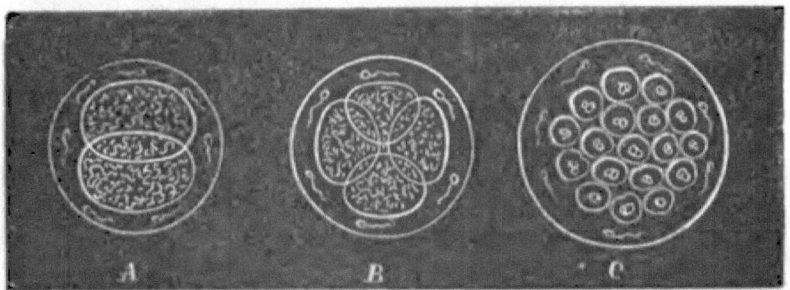

FIG. 20.—Segmentation of Mammalian Egg. A. Division into halves. B. Further
subdivision. C. Mulberry mass, or Morula.

in the vertebrates arrange themselves into a layer lining
the vitelline membrane, on one side of which is a sort of
pouch, or *blastoderm*, consisting of three layers of cells,
the epiblast, the mesoblast, and the hypoblast. The
first of these produces the skin, the middle one the
nervous, muscular, and vascular systems, and the latter
the lining of the intestinal and respiratory organs.

The alimentary canal is at first a straight tube closed
at both ends. As it grows faster than the body it is

6*

thrown into a spiral coil, and at several points it dilates, to form the stomach, etc. The mouth is developed from an infolding of skin. The liver is an outgrcwth from the digestive tube, at first a cluster of cells, then of follicles, and finally a true gland. The lungs first appear as minute buds from the upper part of the aliment- ary canal, or pharynx.

15. The transformation of the cells of the blastoderm into various animal tissues is effected in various ways.

a. An interstitial deposit of formed material may occur in the bioplasm, or cell. Thus oil-globules, pigment, or vacuities may greatly modify the appearance and actions of the cell. The action of tannin, or boracic acid, etc., upon the red blood disks of animals, shows each of them to be really double, having a continuous interstitial sub- stance deposited in each disk. Prof. Brucke, who first investigated this structure, called the parts of the disk respectively, the *zooid* and the *œcoid*, the former being the part which, in the living state, contains also the hæmoglobulin, or red coloring matter.

b. Cells are sometimes found scattered through an intercellular material, the product of cells or of cells transformed and fused together. This intercellular mass may either remain continuous, or split up into fibers. In this way fibrous connective tissue, cartilage, etc., may be formed. (Fig. 21.)

c. The cells which cover surfaces, and through which all interchange between the body and the external world is carried on, are called *epithelial*. They differ in shape, either from mutual pressure or function, some being flat and squamous, (or scaly,) and others columnar.

Some of the latter have cilia, or hair-like projections, whose motions produce a current over the surface.

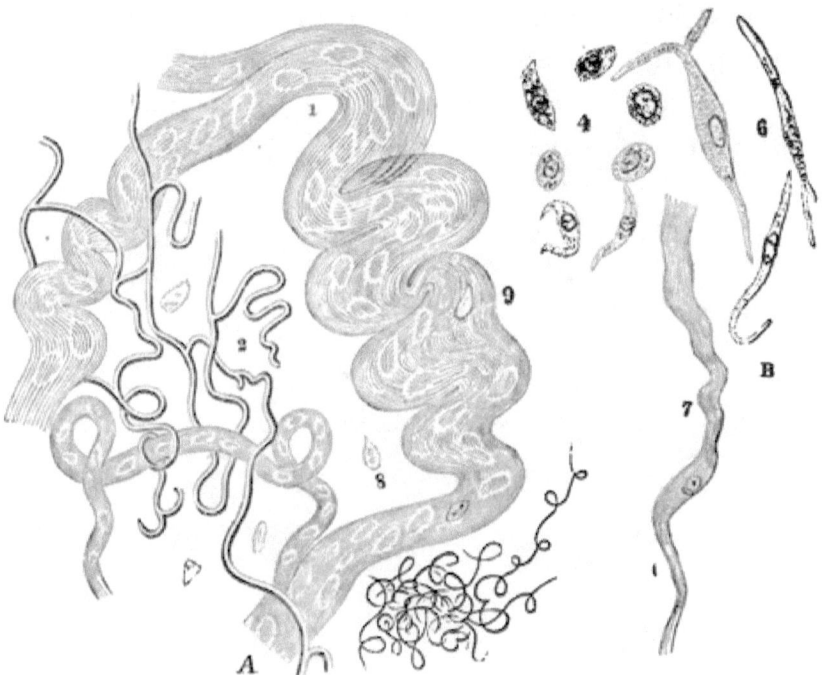

FIG. 21.—Connective-tissue. A. White and yellow fibers. B. Developing Cells of connective tissue.

Thus the skin, or mucous membrane, is not a continuous membrane, but made up of cells, the nuclei of which exhibit the remains of the bioplasm or living matter from which they sprang. (Fig. 22.)

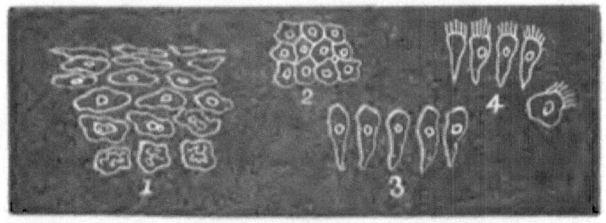

FIG. 22.—Epithelial cells. 1. Squamous epithelium from the skin, showing the change from bioplasm to horny scurf. 2. Tessellated Ep. from serous membrane. 3. Columnar Ep. from intestine. 4. Ciliated Ep. from air-passages.

d. In bone and other hard tissues, as the teeth, the intercellular substance is solidified by salts of lime deposited in a modified form by molecular coalescence. Sec. 3.   In this case the bioplasm, or cell, is limited to certain spaces, or *lacunæ*, and receives nourishment through small canals, or *canaliculi*. (Fig. 23.)

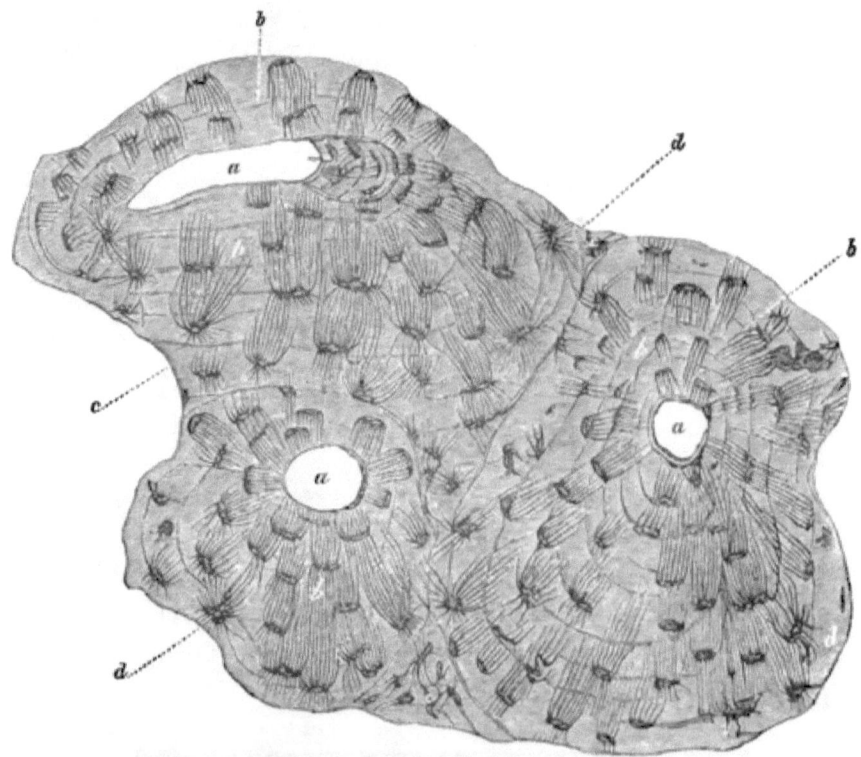

Fig. 23.—Transverse section of a long bone. *a.* Haversian canal. *b.* Concentric laminæ. *c.* Laminæ of connection. *d.* Lacunæ, with their system of tubes.

e. Some fibrous structures may be formed by moving particles of bioplasm, leaving behind them a thread of formed material.   In voluntary muscular fiber this formed material is duplex, and in certain nerve-ganglia the fiber is spirally coiled around another by the forward and rotary motion of the bioplasmic cell.

16. We may consider the living organism, either animal or vegetable, as a building, a workshop, or a laboratory, and in each view the cell, or bioplasm, plays the most important part.

If we regard an organism as a building, the cells are the constituent parts, or building-stones. The most simple forms of life, as we have said, are single cells, while the more complex are composed of myriads of these cells, with the materials produced by them, arranged in various forms, according to the nature of the individual. Thus in the yeast-plant (*Torula*) the cells touch each other at only one or two points, while the wood-cells of higher plants adhere in their entire extent by means of formed material. Vessels, or ducts, are either elongated hollow cells, or are formed by the union of cells. In every structure, except the most primitive, we also find secret chambers and grottoes which we should not previously have suspected; and where strength is needed, provision is made for it by the deposit of hard substance, and by the interlacing of fibers, once cellular, in a most wonderful manner. Even the temple of Solomon, in all its glory, was not more complete in architectural details than the structure of many of our plants and animals. As that temple was said to have been erected without the sound of hammer or saw, so the animated edifice is built silently, story after story, from day to day, until its life-work is accomplished.

Such a structure is a workshop, as well as a building.

There is something in it full of peculiar activity, altogether different from the forces which belong to metals
and stones, or other inorganic bodies.   We call it Life,
and the more we observe its powers the more we shall
be convinced that it is the Master, and not the slave, of
matter, and that the forming power is different from the
thing which is formed.   It makes its own workshop and
its own tools, and compels the physical forces of inorganic nature to assume new and different relations, so as
to serve its own purposes.   It forms its own building-
stones, and elevates them to their places against gravity,
removes such as are in decay and replaces them with
others, and strengthens such parts as are most exposed
to wear or strain.

The organism is also a laboratory.   There Life, as a
subtle Alchemist, sits and transmutes the chemical elements around it into new and useful forms, in a way
which surpasses all our knowledge.   Thus from the same
materials, and under the same conditions of light, heat,
and electricity, one cell will make starch, another fat,
another sugar, albumen, flesh, coloring-matter, acids, or
alkalies ; nay, even in parts of the same cell different
materials may be produced.

17. Every glance into the marvels of organic structure
reveals new wonders.   As in the remote regions of
space we may trace myriads of suns, with nebulous films
and world-islands, which hide from us what is behind
them, so here every step reveals something new and
gives glimpses of something beyond.   The details of
Histology would fill a large volume, and even an ordinary life-time is insufficient to do more than to gather

up a few facts and arrange them in proper relations, yet
the pursuit of knowledge continually brings us nearer to
the fountain of Absolute Truth. To the microscope,
even more than to the telescope, belongs the introduction of the inquirer into the arcana of the universe. If
it does not lead us outward into realms of space, which
exhibit the same relations of scientific and abstract truth
as the world on which we dwell, it leads us inward toward the foundations of our own existence, and shows
that the relations of truth are as perfect in the descend-
ing as in the ascending sphere. If we see not life itself,
we see its first beginnings, and the process of its development. If we see not Nature in her undress, we trace
the elementary warp and woof of her mystic drapery.
From both telescope and microscope alike we learn
that the widening sphere of knowledge is constantly
encircled by the unknown, yet through them we see
above and beneath us a myriad instances of the skill and
providence of a Great Designer, who is God and Father
of all. The living atom shines with truth no less than
the star.

> " Forever singing as they shine,
> The hand that made us is divine."

CHAPTER V.

TYPES OF CONSTRUCTION.

. . . Much less, then, have we any idea of the substance of God. We know him only by his most wise and excellent contrivances of things and final causes ; we admire him for his perfections ; but we reverence a id adore him on account of his dominion ; for we adore him as his servants ; and a god without dominion, providence, and final causes, is nothing else but Fate and Nature. Blind metaphysical necessity, which is certainly the same always and every-where, could produce no variety of things. All that diversity of natural things which we find suited to different times and places could arise from nothing but the ideas and will of a Being neces- sarily existing.—SIR ISAAC NEWTON'S *Principia.*

1. OUR imperfect knowledge of nature must always give a provisional character to our classifications. If they present the knowledge we possess in a useful and compact form, it is all they can be expected to do. Fur- ther knowledge may confirm or overthrow the most per- fectly symmetrical system. Tennyson has well sung :

"Our little systems have their day,
   They have their day and cease to be,
They are but broken lights of thee,
   And thou, O Lord, art more than they."
                                    —*In Memoriam.*

Yet an arrangement may be true although imperfect. We may see plainly the leading outline, while a myriad details may be unknown.

2. In attempting to arrange organic forms it is impos- sible to place them in a single line, like the steps of a ladder, according to structural rank. There are no such

gradations in nature as some imaginations have conceived. There are so many relationships, both of structure and of function, that a single series is out of the question. There are many series, and series, also, within series. Organic forms seem to be placed in radiating groups rather than lines, each group being connected, not with two groups merely, one above and the other below, but with several. Living things are, therefore, best studied in groups, or circles, according to prominent types or representative forms. These groups will, doubtless, be unequal and dissimilar, and will be far from representing the grade of organization ; yet they will be of great use, not only to the memory, but also in indicating the general order of the universe.

3. The unity of organic nature is seen in the similarity of bioplasm, or living matter ; its variety is shown in the multiform arrangements of structure in living beings. That all this variety can be intelligently connected together in a few comprehensive groups, exhibiting plans of structure, is proof positive of the intelligence of the creative power. Agassiz has well said, " If these classifications are not mere inventions, if they are not an attempt to classify for our own convenience the objects we study, then they are thoughts which, whether we detect them or not, are expressed in Nature—then Nature is the work of thought, the production of intelligence, carried out according to plan, therefore premeditated—and in our study of natural objects we are approaching the thoughts of a Creator, reading his conceptions, interpreting a system that is his and not ours."

4. *Types* are comprehensive natural groups of living

7

forms, founded on plans of structure or structural ideas. *Classes* comprise all forms which agree simply in a special modification of the type to which they belong. The type represents the plan, but there may be several ways of executing the plan, and these ways illustrate the classes. In human works of art " there are certain architectonic types, including edifices of different materials, with an infinite variety of architectural details and external ornaments ; but the flat roof and the colonnade are typical of all Grecian temples, whether built of marble or granite or wood, whether Doric or Ionic or Corinthian, whether simple and massive or light and ornamental ; and, in like manner, the steep roof and pointed arch are the typical characters of all Gothic cathedrals, whatever be the material or the details. The architectural conception remains the same in all its essential elements, however the more superficial features vary. Such relations as these edifices bear to the architectural idea that includes them all, do classes bear to the primary divisions," or types.* Thus Fishes, Amphibians, Reptiles, Birds, and Mammals are classes under the Vertebrate type of animal life.

An *Order* is a group of families, or genera, related to one another by a common structure. Thus Cats, Dogs, Hyenas, and Bears are linked together, since their teeth, stomachs, and claws show the carnivorous habits of the order Carnivora.

A *Family*, or *Tribe*, does not allude to the progeny of a known stock, but refers to a group of genera having similarity of form. The term was first introduced into

* Agassiz, " Methods of Study."

Botany in France, in connection with what is called the natural system of classification. To prevent confusion, the similarity of form determining families should be based on structure and not mere resemblance.

A *Genus* is a group of species having the same essential structure. Thus the allied species, Cat, Tiger, and Lion, belong to one genus.

A *Species* is the smallest group of individuals which can be defined by several constant characteristics. They are so alike that it is possible for them to have descended from one pair. A cross between two species, as the Horse and Ass, is called a *hybrid;* as the Mule.

Individuals are the units of organic life. A complete animate existence is an individual, whether separate, as man, or living in a community, as the Coral. When two or more individuals differ by a single peculiarity, such as color, or outline, or size, one is called a *variety* of the other, as the races of Men and breeds of Cattle. A cross between distinct races is called *mongrel.*

Vegetables and animals are separated from each other under the term *kingdom*, and the types of structure in each kingdom are called *sub-kingdoms*. Thus in the animal kingdom we have the sub-kingdoms of Vertebrates, Radiates, etc.

There are no such *things* as genus, species, order, class etc. They are but abstract terms, expressing relation to a plan, or the harmony of intelligent design which presides over all things.

5. A real type, or plan, includes all those individuals, species, etc., which are similar in character. But it is not always easy to determine similarity of character.

From the earliest times of history down to Cuvier, naturalists were in the habit of regarding similarity of external form and evident purpose as indicating analogies, and so far as functional design is concerned, the principle may be considered right. But purpose and plan for a purpose are different, and modern science seeks for its types in the characters of internal structure and development.

6. Parts, or organs, having similar origin and development, and therefore the same essential structure, are called *homologous;* while those which are anatomically different, though corresponding in use, are called *analogous.* Thus in the vegetable kingdom the tendril of the Vine, which is a transformation of the flower-stalk; that of the Pea, which is a prolongation of the leaf-stalk; that of *Gloriosa*, which is the point of the leaf itself; and that of *Strophanthus*, which is the point of the petal; are all analogous, but not homologous. The arms of Man, the fore-legs of a Horse, the paddles of a Whale, the wings of a Bird, the front flippers of a Turtle, and the pectoral fins of a Fish, are homologous but not analogous. The wings of the Bird, Flying Squirrel, and Bat are not homologous, since that of the first is developed from the fore-limb only, that of the Squirrel is an extension of the skin between the fore and hind limbs, and that of the Bat is a membrane between the fingers and down the side to the tail. The air-bladder of a Fish is homologous with a lung, but analogous to the air-chamber of the Nautilus. In the functional analogies, perhaps more evidently than in the structural homologies, we trace evidence of purpose, or design. "Blind metaphysical

necessity," as Newton called Fate and Nature without God, could certainly produce no such "variety of things" as we see here, while the unity pervading the functional character of the different organs is plain enough proof of their being the work of the same Artisan.

Various functions are attained by a modification of similar structure. Thus the simplest plant differs from the most complex principally in this—that the whole external surface of the former participates equally in all the operations which connect it with the external world, as those of Absorption, Exhalation, and Respiration, while in the latter these functions are confined to certain parts of the surface. So in the highest animals, the organs adapted to the functions of Absorption, Exhalation, Respiration, Secretion, and Reproduction, are all composed essentially of a membrane which is a prolongation of the general surface, while this general surface is the sole instrument for the performance of these functions in the lowest animals, and shows no special adaptation for one or another of them. So that it may be expressed as a general truth of Biology, that "throughout all animate Creation, the functional character of the organs which all possess in common, remains the same; while the mode in which that character is manifested varies with the general plan upon which the being is constructed."[*]

In all living things the attainment of function is the cause of modification of structure. This gives evidence of Creative plan, or design, in direct opposition to the theory of gradual evolution of structure, and is proof also

[*] Carpenter's "General and Comparative Physiology."

7*

of the essential differences between living beings, since the plan of structure varies for attaining similar purpose.

7. Cuvier proposed four primary divisions of the animal kingdom, because, he said, they are constructed on four different plans. These plans may be briefly stated as follows: "In the *Vertebrates* there is a vertebral column terminating in a prominent head; this column has an arch above and an arch below, forming a double internal cavity. The parts are symmetrically arranged on either side of the longitudinal axis of the body. In the *Mollusks*, also, the parts are arranged according to a bilateral symmetry on either side of the body, but the body has but one cavity, and is a soft, concentrated mass, without a distinct individualization of parts. In the *Articulates* there is but one cavity, and the parts are here again arranged on either side of the longitudinal axis, but in these animals the whole body is divided from end to end in transverse rings or joints movable upon each other. In the *Radiates* we lose sight of the bilateral symmetry so prevalent in the other three, except as a very subordinate element of structure; the plan of this lowest type is an organic sphere, in which all parts bear definite relations to a vertical axis." * Leuckart proposed to subdivide the Radiates into two groups; the Cœlenterata, including Polyps and Acalephs, or Jelly-fishes—and Echinoderms, including Star-fishes, Sea-Urchins, and Holothurians, but Agassiz shows that the differences between them are not differences in the plan, but merely a difference in the execution of the plan, since both are equally radiate in structure.

* Agassiz, "Methods of Study."

By this radial symmetry we are conducted toward the Vegetable Kingdom. Thus in the higher *Fungi* the disposition of organs is as radiate as in Radiated animals. In *Mosses* and *Ferns* there is a spiral arrangement of leaves around the axis, which may be considered the regular law of growth in the higher plants, although sometimes obscured by special modifications.

8. It is a popular error, fostered by the assertions of certain Monistic writers, that the higher animals pass through all the phases of lower life. This false notion is based upon too strict an interpretation of Von Baer's generalization in Embryology, that "a heterogeneous or special structure arises by gradual change out of one more homogeneous or general." Every division of the Animal Kingdom has its characteristic method of developing. "The Vertebrate arises from the egg differently from the Articulate; the Articulate differently from the Mollusk; the Mollusk differently from the Radiate."* "Every grand group early shows that it has a peculiar type of construction. Every egg is from the first impressed with the power of developing in one direction only, and never does it lose its fundamental characters. The germ of the Bee is divided into segments, showing that it belongs to the Articulates; the germ of the Lion has the primitive stripe—the mark of the coming Vertebrate. The blastodermic layer of the Vertebrate egg rolls up into two tubes—one to hold the viscera, the other to contain the nervous cord; while that of the Invertebrate egg forms only one such tubular division. The features which determine the subkingdom to which an animal belongs

* Agassiz.

are first developed, then the characters revealing its class."* Dr. Carpenter says: "The human embryo is never comparable with a Fish, a Reptile, or a Bird, much less with an Insect or a Mollusk. However close may be the resemblance between the embryo of Man and the embryo Fish, there is no real correspondence between the embryo of Man and the completed Fish. Each germ has a certain capacity of development peculiar to itself, since *like produces like.*"

9. To attain a true knowledge of the order of creation, or of the types of structure among organic forms, it is necessary not only to consider internal construction and relationships, and the process of embryonic development, but also to trace the life-history of each, and especially the metamorphoses to which they may be subject at various periods. Among the lower Fungi there is a kind of *polymorphism* (*polys*, many; *morpha*, form) frequent, by which several forms may be developed by spores, or seeds, which have identically the same origin. Few animals come forth from the egg in perfect condition. The embryonic Star-fish has a long body, with six arms on a side, from one end of which the young Star-fish is budded off. Soon the twelve-armed body dies, and the young animal is of age. Most Insects undergo complete change of form, a *metamorphosis; i.e.,* exhibit four distinct stages of existence—egg, larva, pupa, and imago. Among the vertebrates the most common and most remarkable transformation is that of the Frog. It is first, after hatching from the egg, a tadpole, with a tail, but no legs, gills instead of lungs, a heart like that

* Orton's "Comparative Zoology."

of a fish, a beak for eating vegetable food, and a spiral intestine to digest it. As it matures, the hind legs show themselves, then the front pair, the beak falls off, the tail and gills waste, lungs are formed, the digestive apparatus is changed to suit an animal diet, the heart is altered to the Reptilian type by the addition of another auricle—in fact, skin, muscles, nerves, bones, and blood-vessels vanish, being absorbed atom by atom, and a new set is substituted.*

10. With the caution referred to in Sec. 1 we may present an outline of the types of living forms.

The most general and comprehensive type is that of bioplasm, or living matter, described in Chap. II, and characteristic of both animals and vegetables. The next most comprehensive type of structure is that of Vegetable forms in which the bioplasm is invested, or, as it were, imprisoned, in each of its component cells by a sac of cellulose, or some analogous compound, (Chap. IV, Sec. 5,) and whose most characteristic work, or peculiarity, is its power of manufacturing albuminoid matter out of simpler chemical elements. In Animal forms there is no such cellulose investment, nor can they make albuminoid matter from inorganic elements.

In the Vegetable Kingdom we may arrange organic forms under the following general divisions, or principal types:

1.) PROTOPHYTES, or simplest vegetable forms, answering to the *Unicellular Algæ*.

2.) THALLOGENS, which are a mere expansion of cellular tissue, without complete distinction between stem,

* Orton.

root, and leaves. These include *Fungi*, *Algæ*, and *Lichens*.

3.) ACROGENS. Plants which grow in height and not in diameter. *Liverworts*, *Mosses*, and *Ferns*.

4.) ENDOGENS. Vascular plants, in which the wood and cellular tissue are mixed throughout, without distinct annual layers. The seed has but a single lobe, or *cotyledon*.

5.) EXOGENS. Vascular plants having distinct annual layers of woody fibers, and radiations of tissue from the medulla to the bark. The embryo has two seed-lobes, or cotyledons.

In the Animal Kingdom we have the following typical forms, or subkingdoms:

I. PROTOZOA. Simplest animal forms, being composed of bioplasmic jelly. *Monera*, *Gregarina*, *Rhizopods*, *Infusoria*, and *Sponges*.

II. RADIATA. Radiate animals, which are subdivided into—1. CŒLENTERATA, with distinct body-cavity, tentacles, and nettling thread-cells. *Hydrozoa*, *Anthozoa*, *Ctenophora*. 2. ECHINODERMATA, with distinct alimentary canal and nervous ring. *Crinoids*, *Asteroids*, *Holothurians*, *Echinoids*.

III. MOLLUSCA. Soft unsymmetric animals. Digestive system developed. Nervous system irregular. *Polyzoans*, *Tunicates*, *Brachiopods*, *Lamellibranchiates*, *Gasteropods*, *Cephalopods*.

IV. ARTICULATA. Nervous ventral cord double. Limbs on same side as nerve-cords. *Annelids*, *Crustaceans*, *Arachnoids*, *Myriapods*, *Insects*.

V. VERTEBRATA. Double nervous system; one on

upper side of alimentary canal, the other spinal; limbs opposite nerves. *Fishes, Amphibians, Reptiles, Birds, Mammals, Man.*

11. In the Frontispiece the characteristic features of biological types are represented. In the outline section of each of the four types of higher animals the large shaded spot shows the alimentary canal, the dark spot the position of the heart, and the open rings the nervous system. A diagram of the latter also accompanies each of those types.

CHAPTER VI.

PROTOPHYTES.

Let no presuming impious railer tax
Creative Wisdom, as if aught was formed
In vain, or not for admirable ends.
Shall little haughty ignorance pronounce
His works unwise, of which the smallest part
Exceeds the narrow vision of her mind?
—THOMSON.

1. VEGETABLE structure has been already character-ized as bioplasm imprisoned, or invested with a cell-wall of cellulose. In some of the simplest forms, or *Proto-phytes*, each cell is separate from the rest, others form masses of cells in a sort of gelatinous or slimy invest-ment, while other forms exhibit a definite adhesion be-tween the cells, so as to prefigure a regular plant-like structure, although each cell is a repetition of its parent-cell, and is capable of living apart.

2. The life-history of simplest Protophyte is exem-plified in the *Pal-moglœa macrococ-ca*, (Fig. 24;) a sort of green scum or slime, growing on damp stones, etc. The microscope shows this to con-sist of a multitude

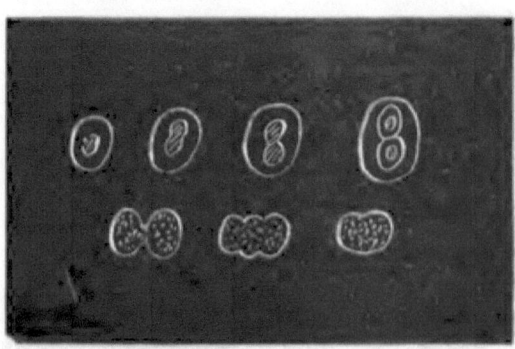

FIG. 24.—Development of Palmoglœa macrococca.

of green cells, each surrounded by a gelatinous envelope, and sometimes a *nucleus*, or more solid aggregation, which is considered the center of vital activity, is seen in the cell. The green particles, or granules, which fill the cells, are formed material called *chlorophyll*. Throughout the vegetable kingdom the presence of chlorophyll is necessary to enable the plant, under the stimulus of bright sunlight, to break up carbonic acid, evolve the oxygen, and appropriate carbon as food. In the absence of sunlight all plants become oxidized, and evolve carbonic acid. The cells of the Palmoglœa multiply by binary subdivision, or fission. (Chap. III, Sec. 12.) This multiplication is an act of growth, and differs from similar self-division in the higher plants by the purpose manifested, and the plan for a purpose, seen in the "differentiation" of cells in the higher orders for the production of special organs.

In these lowly plants there is a process similar to the plan of reproduction in the more complex forms. A pair of cells will sometimes reunite, or fuse together, first by means of a narrow bridge, and then a larger mass, and finally a complete fusion. The mass is termed a *Sporc*, (from the Greek *spora*, a seed,) and is the primitive cell of a new generation formed by fission.

3. In a form allied to the above, the *Protococcus pluvialis*, (Fig. 25,) not uncommon in rain-water, a somewhat greater variety of conditions has been seen. It is found *still*, or quiescent, and *motile*. In the first form the bioplasm is surrounded with a wall of cellulose, and filled with granules of green or red chlorophyll. These still cells multiply by self-division, each producing two, four,

8

eight, or sixteen new cells. The new cells are motile, having each two long vibratile filaments or *cilia*. They may

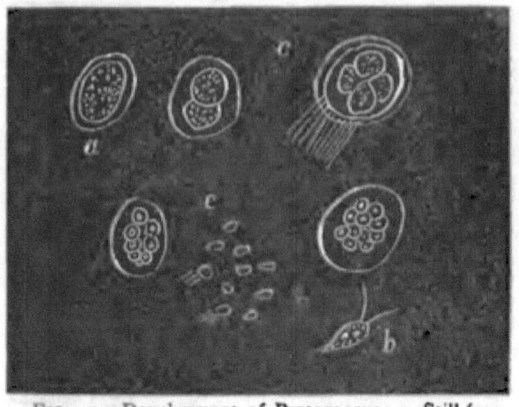

be seen swimming, tumbling, or rotating in the water. At times they are surrounded by a sac, which may be at some distance from the bioplasm. The motile cells may also multiply by subdivision, and

FIG. 25.—Development of Protococcus: *a.* Still form. *b.* Motile form. *c.* Self-division and zoospores.

in some cases the minute primitive cells, when set free, have very active movements, and rank as *Zoospores*, (living spores,) or *Micro-gonidia*, (small angular particles, from division of the bioplasm.) The varieties connected with the history of this single plant have been sometimes described as distinct species, and even genera of Animalcules, because of their shape and motions.

4. The family of *Palmellaceæ*, to which the forms referred to belong, contain some kinds of singular interest. The "Red Snow," which sometimes colors extensive Alpine or Arctic tracts, is composed of myriads of Protococcus cells, in a quiescent state, with the chlorophyll of a red color. The *Palmella cruenta*, or "Gory Dew," appears sometimes as tough gelatinous masses of the color of coagulated blood, and extends over a considerable area. In this way we may account for showers of flesh, blood, etc., which are often regarded as bad omens by the ignorant.

5. The family *Volvocinæ* has been long considered
of singular beauty and interest to the microscopist.
The *Volvox globator* (Fig. 26) was described by Leeu-
wenhöek about
one hundred and
fifty years ago,
and from its shape
and rolling mo-
tion was called the
globe animalcule,
but its vegetable
character is now
generally admit-
ted.   It is about
one thirtieth of

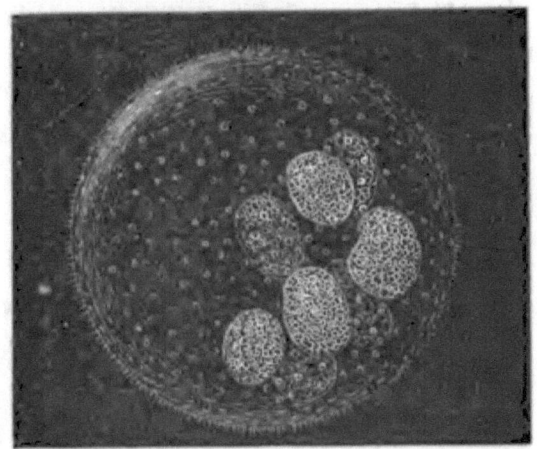

FIG. 26.—Volvox globator.

an inch in diameter, and appears to the unassisted eye
to be a little green speck moving slowly through the
water.   On examining with the microscope the Volvox
is seen to be a pellucid sphere studded with minute
green spots, connected together by threads.   From each
of these spots proceed two cilia, so that the entire surface of the globe is beset with vibratile filaments, to whose
combined action its rolling motion is due.   Within the
globe may be generally seen from six to twenty other
globes, of varying sizes, which are set free by the bursting of the parent globe.   Sometimes a third generation
may be seen within the secondary spheres.   Careful observation of the development of the Volvox has shown
that the ciliated cells referred to above, analogous to the
zoospores of Protococcus, sometimes appear like moving
Amœbæ. (Chap. II, Sec. 2.)   This is not an uncommon

phenomenon among Protophytes, and shows that the bioplasm of the vegetable and animal cell have similar properties.

6. Dr. Carpenter recommends those who wish to study the development of "zoospores," and other phenomena of Protophytes, to have recourse to the little plant called *Achyla prolifera*, which grows parasitically upon the bodies of dead flies in water, etc. The naked eye perceives it as tufts of minute colorless filaments, which the microscope shows to be long tubes containing granular bioplasm, which exhibits the motion called Cyclosis. (Chap. IV, Sec. 6.) After about thirty-six hours the bioplasm accumulates in the dilated ends of the filaments, and its endochrome, or granular coloring matter, breaks up into distinct masses, each of

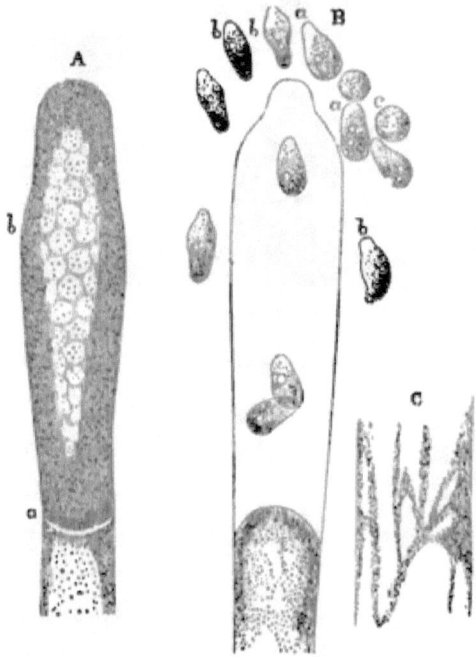

FIG. 27.—Development of *Achyla prolifera :* A. Dilated extremity of a filament, *b*, separated from the rest by a partition, *a*, and containing young cells in progress of formation. B. Conceptacle discharging itself, and setting free young cells, *a, b, c.* C. Portion of filament, showing the course of the circulation of granular protoplasm.

which becomes a zoospore, or "motile gonidium," with cilia, and is set free by rupture of the wall of the parent cell. (Fig. 27.)

7. The family *Desmidiaceæ* consists of minute Proto-

phytes of a grass-green color, growing in fresh water. The cells are generally independent, but in some species remain adherent one to another so as to form a filament.

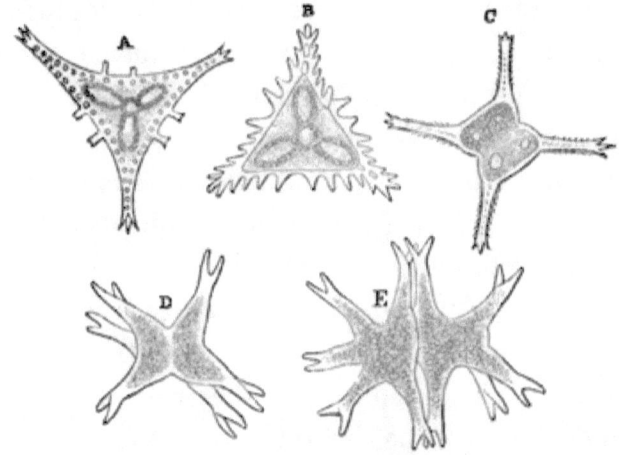

FIG. 28.—Various species of *Staurastrum:* A. *Staurastrum vestitum*. B. *Staurastrum aculeatum;* C. *Staurastrum paradoxum;* D. E. *Staurastrum brachiatum.*

Some species have spiny projections of the outer coat, which is of a horny consistence, as in Fig. 28. Others are notched on the sides; some, as the *Closterium*, (Fig. 29,)

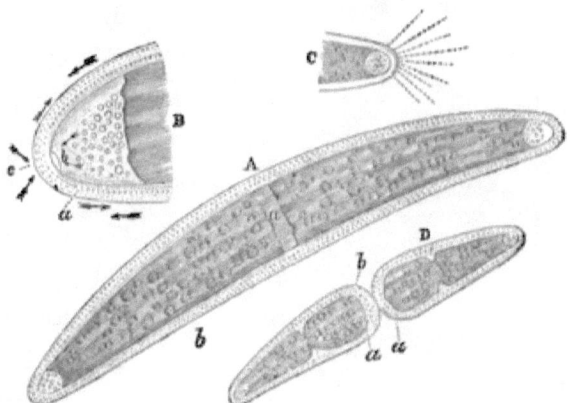

FIG. 29.—Economy of *Closterium lunula:* A. Frond showing central separation at *a*, in which large globules, *b*, are not seen. B. One extremity enlarged, showing at *a* the double row of cilia, at *b* the internal current, and at *c* the external current. C. External jet produced by pressure on the frond. D. Frond in a state of self-division.

are smooth. In the latter a circulation of fluid may be seen between the cellulose coat and the "primordial utricle." (Chap. IV, Sec. 6.) Some consider this circulation to be caused by cilia, but it is rather doubtful. We are inclined to regard it as an exhibition of the molecular motion of bioplasm already described. Many of the Desmids multiply by subdivision, but the plan is modified so as to maintain the symmetry characteristic of the tribe. At other times multiplication takes place by the subdivision of the endochrome into granular particles, or "gonidia," set free by rupture of the cell-wall.

The process of conjugation differs from that of Palmoglæa, since each cell has a firm external envelope, which cannot coalesce with another. In *Cosmarium*, (Fig. 30,) for example, the conjugating cells become deeply cleft and separate, so that the contents pour out freely, at first without a protecting membrane. At length it acquires an envelope, and becomes a *sporangium*, or spore-case, of reddish-brown tint. It is covered with spines, and greatly resembles certain fossil forms found in flint called *Xanthidia*. The Closteria conjugate after a somewhat similar manner, and it is not uncommon to find a pair in this condition, but

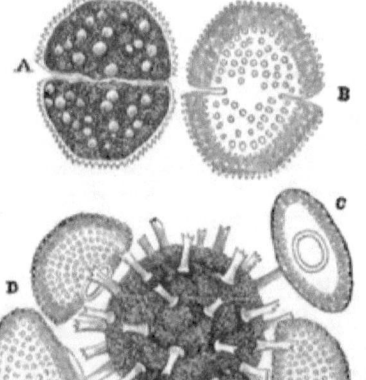

FIG. 30.—*Cosmarium botrytis:* A. Mature frond. B. Empty frond. C. Transverse view. D. Sporangium, with empty fronds.

their sporangia are smooth instead of tuberculated or spiny.

8. The families of Algæ, called *Oscillatoriaceæ*, *Nosto-chaceæ*, *Confervaceæ*, and *Conjugateæ*, may all be considered as Protophytes, but a brief description only can be given here. The structure is generally microscopic.

The *Oscillatoria* are tubular filaments with partial subdivisions, formed by the elongation of their primordial cells, occurring in fresh and salt water, and on damp ground. They have very curious movements, sometimes swaying like a pendulum, and at others bending at the extremity from one side to another, or moving straight onward.

Nostocs are beaded filaments lying in masses of greenish gelatinous matter. As the jelly forms rapidly in damp weather, they have been termed "fallen stars." The alchemists often refer to this substance, and it enters into many of their recipes for the pretended transmutation of metals.

The *Confervæ* may be found in almost every pond or ditch, but are especially abundant in running water. They constitute the greater part of those green threads which are found in streams, or near the sea-shore. Each thread is a long cylinder, in which the endochrome, of a green, brown, or purplish hue, is either distributed uniformly through the cell, or arranged in a net-work, or spiral form. It increases by binary subdivision in the terminal cell, as well as by zoospores produced within the cells.

The family *Conjugateæ* is so called because the filaments are so constantly yoked together. They are

generally found in still water.   In an early stage of
growth the endochrome is diffused through their cavities, but after a time arranges itself in regular spirals.
Then adjacent cells put forth protuberances which coalesce, and a passage is formed between the cells.   The
endochrome of one cell passes over into the cavity of the
other, forming a sporangium, or spore-case.   (Fig. 31.)

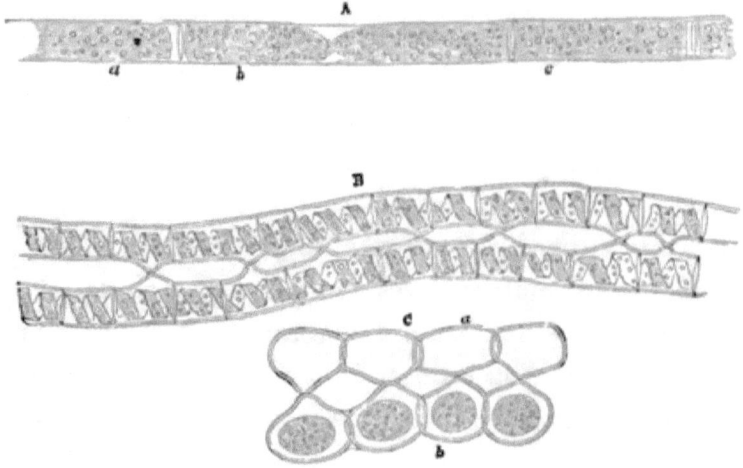

FIG. 31.—Various stages of the history of *Zygnema quininum:* A. Three cells, *a, b, c*,
of a young filament, of which *b* is undergoing subdivision : B. Two filaments in the first
stage of conjugation, showing the spiral disposition of their endochromes, and the protuberances from the conjugating cells.   C. Completion of the act of conjugation, the endochromes of the cells of the filament *a* having entirely passed over to those of filament *b*,
in which the sporangia are formed.

9. The most beautiful and interesting unicellular forms,
now generally conceded to be vegetable, are found among
the *Diatomaceæ*.   Their motions caused many to regard
them as animals, but naturalists now agree in calling
them Protophytes.   They are called Diatoms because
of their extreme brittleness and the ease with which
a chain of them may be broken into its component cells.
Like the Desmids, they are simple cells containing endo-

chrome, with a firm external covering. In the Diatoms, however, this envelope is consolidated by silex, or flinty matter, sometimes also by iron. The silicious envelope of each "frustule," or cell, is covered with most elaborate and beautiful marking, and consists of two valves, or plates, closely applied to each other, like the valves of a Mussel, along a suture, or line of contact. If the valve is hemispherical, the cavity will be globular; if a segment of a sphere, the cavity will be lenticular. Sometimes the central part is flattened, and the sides turned up, so that the valve resembles the cover of a pill-box, in which case the cavity will be cylindrical. Then, again, the valve may be square, triangular, round, heart-shaped, boat-shaped, etc. In many species of Diatoms the markings are so minute that they can only be made out with the highest powers of the microscope; in others a very moderate power suffices to exhibit the lines and dots in patterns which rival the most elaborate works of art. (Figs. 32, 33, 34.)

In the living state Diatoms are found abundantly in every pond, rivulet, ocean, and rock-pool. In some parts of the world they form immense deposits.

A mud bank in Victoria Land, 400 miles long and 120 broad, is composed of silicious valves of Diatoms. In Sweden and Norway, under the name of *bergh-mehl*, they are used for mixing with flour for bread in seasons of scarcity. Under the cities of Richmond and Petersburgh, Virginia, is a deposit twenty feet thick, while the polishing slate of Bilin contains Diatoms so small that in a single cubic inch 40,000,000,000,000 (forty trillions) are found.

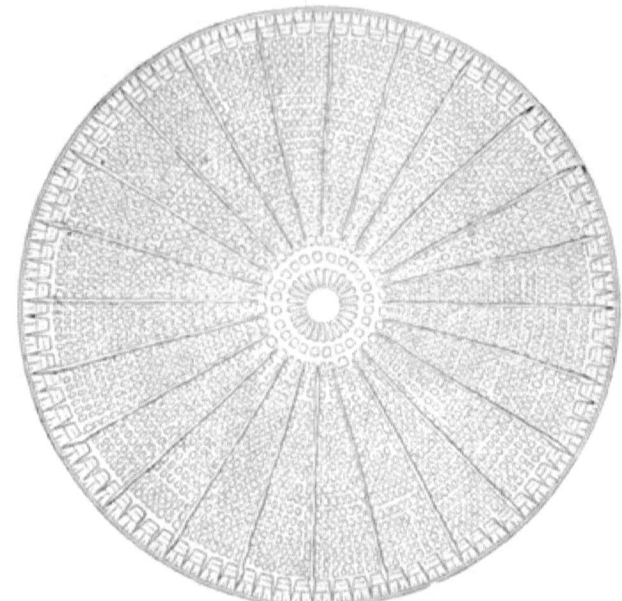

FIG 32.—Arachnoidiscus Ehrenbergii.

The *Arachnoidiscus Ehrenbergii,* (Fig. 32,) is common on the coast of California, attached to sea-weed.   Its general appearance is that of a glassy pill-box. The figure shows one of the ends, or frustules, covered with delicate tracery like a spider's web, by which the genus gets its name, from *arachné,* a spider, and *discus,* a disk.

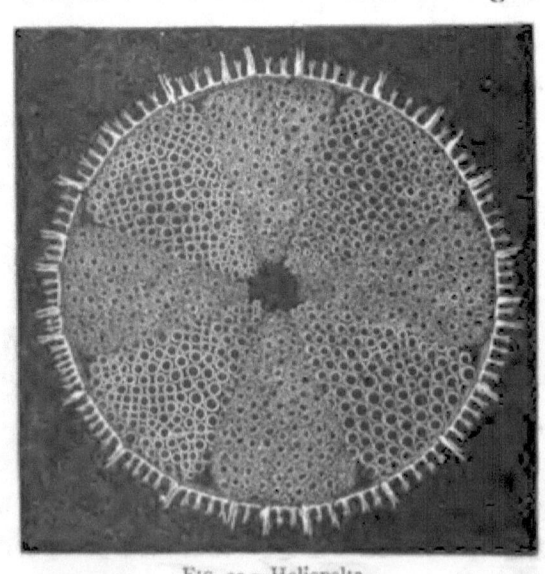

FIG. 33.—Heliopelta.

The *Heliopelta*, sun-shield, (from *helios*, the sun, and *pelta*, a shield, (Fig. 33,) is a most beautiful disk, whose markings form a regular star, the number of whose rays determine the species.

The *Diatoma vulgare*, (A., Fig. 34,) is a quite common

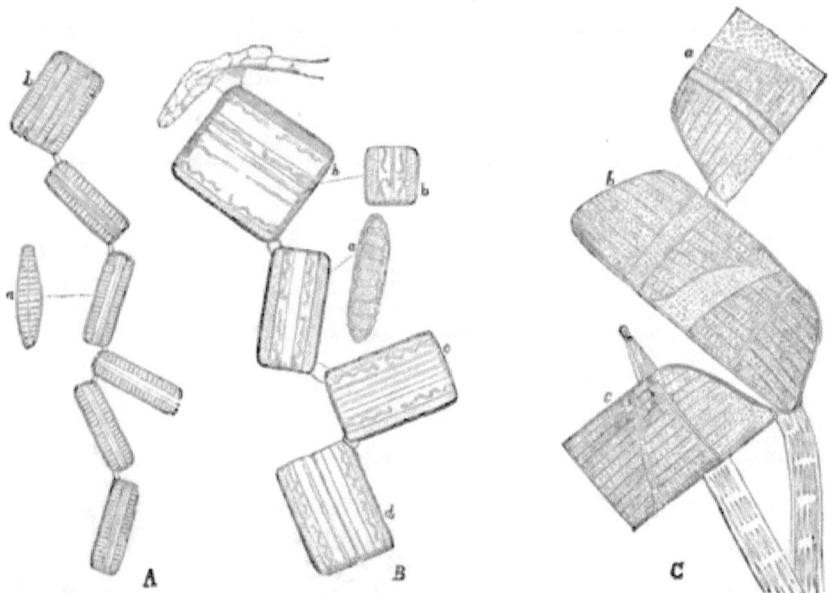

FIG. 34.—A. *Diatoma vulgare:* a. Side view of frustule ; b. Frustule undergoing self-division. B. *Grammatophora serpentina.* a. Front and side views of single frustule. b. b. Front and end views of divided frustule. c. A frustule about to undergo self-division. d. A frustule completely divided. C. *Isthmia nervosa.*

form. The frustules often hang together, forming zig-zag chains by rapid self-division.

Some species of *Grammataphora* have delicate striæ on the borders of each valve, which are used as tests of microscopic excellence.

The *Isthmia nervosa* has a remarkable areolated structure, which will repay careful examination. In its growth two cells form within the valves, and, as they

enlarge, break forth, but the silicious hoop which joined
the new frustules to the old one remains attached for a
time round one of them, causing some to appear truncated instead of round.

The genus *Navicula* is so called from its resemblance
to a boat or little ship, (*naus*, a ship.) They are found
both living and in a fossil state. Some are striped longitudinally, and some transversely; some are shaped like
an old-fashioned letter S, as the *Pleurosigma*, in which
the striae are resolved by a highly magnifying power into
hexagonal dots.

10. In the Protophytes we see the endowments of
simple vegetable cells. A piece of bioplasm, or living
tissue, transforms its outer layer into cellulose, and
forms chlorophyll or starch in its interior, absorbing new
pabulum continually, and casting off the old effete atoms.
The relationship of each family is seen by these functions common to all. Each species of each family has,
however, its own peculiarities, which distinguish it from
all others. The Protococci remain rounded cells, but
the Oscillatoria, Confervaceæ, etc., have an instinct for
elongation, so that they become tubular, and for distrib-
uting endochrome in characteristic spiral patterns, varying in each species, while the Diatoms appropriate silica
from their pabulum to harden the cellulose envelope,
and arrange it in their frustules atom by atom, each
species after its own pattern, and all with marvelous
regularity and beauty. The Monistic theory of the
universe has no satisfactory reason to give for the existence of such varied tendencies. Schleiden has well said,
" We do, indeed, see into the mechanism of the puppet;

but who holds the strings and directs all its motions to *One Purpose?* Here closes the office of the naturalist; he turns from the world of space and lifeless matter upward to where, in holy anticipation, we seek the Ruler of worlds." *

* Schlieden's " Poetry of the Vegetable World."

9

CHAPTER VII.

THALLOGENS.

Thus Nature varies ; man and brutal beast,
And herbage gay, and silver fishes mute,
And all the tribes of heaven, o'er many a sea,
Through many a grove that wing, or urge their song
Near many a bank of fountain, lake, or rill,
Search where thou wilt, each differs in his kind,
In form, in figure differs.—LUCRETIUS.

1. THALLUS-PLANTS, called also Thallogens, or Thallophytes, (from *thallos*, a frond, or green leaf; *genein*, to produce, *phyton*, a plant,) have no true vascular system, but are composed of cells of various sizes, forming membranous expansions, or filaments more or less simple, branched, or interlacing. They differ from Protophytes by the more intimate union of cells in their structure. In some of the Protophytes there is an adhesion of the cells by a fusion of their gelatinous investment, yet each cell is a repetition of the former one, and is capable of living independently if detached, so that each answers to the designation of a unicellular, or single-celled plant. In the Thallogens the cells are not only closely united, but exhibit a differentiation in structure or function, and a relation of mutual dependence, constituting each plant (not each cell) an individual.

2. The higher *Algæ*, or Sea-weeds, the *Lichens*, and the *Fungi* may be regarded as Thallophytes, although

some species may present many points of resemblance with the simple Protophytes.

3. The ALGÆ, or Sea-weeds, have been divided into three orders, the Red, the Olive, and the Green Sea-weeds—Rhodospermeæ, Melanospermeæ, and Chlorospermeæ. The latter order generally includes the Confervoid and other families which we have considered as unicellular plants. When we examine the higher Sea-weeds, we find a certain foreshadowing of the distinction between Root, Stem, and Leaf, which is characteristic of still higher types. This sort of unconscious prophecy of higher forms to come is by no means uncommon in other classes both of plants and animals, and affords another proof that living things have been formed on an intelligent plan. In the Sea-weeds, however, the apparent distinction of stem and root serves for little else than the mechanical attachment of the plant. There is no departure from the cellular type, the only modification being that the layers of cells are of different sizes and shapes.

4. The Olive Sea-weeds, or Fucoids, (*Melanospermæ*,) often grow to a considerable size, attached by sucker-like roots to the rocks, and, in some cases, buoyed up by air-bladders. Others are parasitical. The fructification of many species in this group is sexual. In the common bladder-wrack, *Fucus vesiculosus*, the reproductive organs are on different plants, but in other species, as *Fucus senatus*, they are both together, the one olive-green, the other orange-yellow. The " receptacles " are at the extremities of the fronds, and may be known to be mature by each discharging little gelatinous masses adhering

round its orifice.   If now a section is made through it,
it will be seen to be a cavity lined with filaments, some
of which project through the pore.
The filaments, or *an-
theridia*, are chains
of cells containing
antherozoids. These
are yellow oval bod-
ies, with two thread-
like appendages,
which, when liber-
ated by the bursting
of the cell, have a
spontaneous and rap-
id motion around the
sporangia, (or parent
cells of the germs,)
which they fecun-
date.   These sporan-
gia are pear-shaped
bodies lying near the
walls of the cavity, or
receptacle, and each

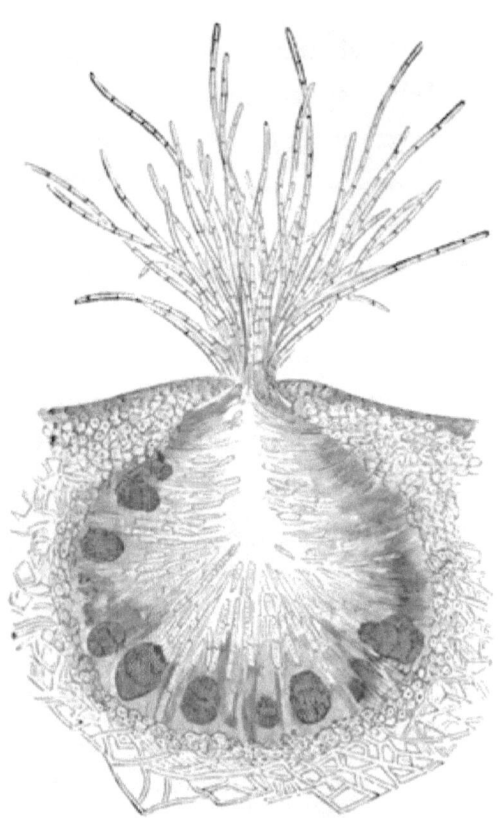

FIG. 35.—Section of receptacle of *Fucus platycarpus*,
lined with filaments containing antheridial cells and
sporangia.

one produces, by fission, a cluster of eight cells, or octo-
spores.   (Fig. 35.)

5. Among the red Sea-weeds, or *Rhodospermeæ*, are
various simple but beautiful forms, eagerly sought for by
sea-side collectors for albums.   They live generally in
deeper water than other sea-weeds, and show their rich-
est tints when growing in the shade.   The genus *Poly-*

siphonia contains many species, some small and delicate, or long and filmy, and of various tints of brown or violet. The fronds are thread-like and jointed; the joints striped, since the stem is composed of parallel tubes or siphons, from whence its name, (*poly*, many; *siphon*, tube.) The fructification is twofold, on distinct plants: 1.) *Ceramidia*, or urn-shaped cells containing pear-shaped spores; 2.) *Tetraspores*, or groups of four spores, imbedded in swollen branchlets. The genus *Ceramium* is thread-like, jointed, branched, with repeated forkings. The tips of the filaments are always forked, and often curl toward each other. The fruit is of two kinds: 1.) Berries, or capsules, containing seeds, and called *favellæ*. 2.) Tetraspores, or groups of four seeds, immersed in the substance of the branch, and surrounding it in a whorl. Another beautiful and not uncommon genus, found at low-water mark, or cast up after a storm, is *Ptilota*, (from a Greek word signifying "pinnated.") It has many small branches, or pinnæ, and these again are cut into smaller divisions, or pinnula. At the top of the latter is the fructification, consisting of minute capsules, or *favellæ*. Some plants also contain tetraspores. *Corallines* are a family of red Sea-weeds whose tissue is consolidated by calcareous deposit. The arrangement of tetraspores in the red Algæ is illustrated by Fig. 36.

6. The class of LICHENS consists of cellular plants of very simple structure. They form irregular patches, generally dry, upon stones, trees, etc., which they decorate with various colors. They are found in all climates. Some are used in medicine, as the Iceland Moss, (*Cetraria Islandica ;*) others, as the Orchil, produce a valuable

9*

dye, and one species, *Leonora esculenta*, found in the Desert of Tartary, seems to fall from the sky as a miraculous manna.   Men and beasts may be nourished on it.

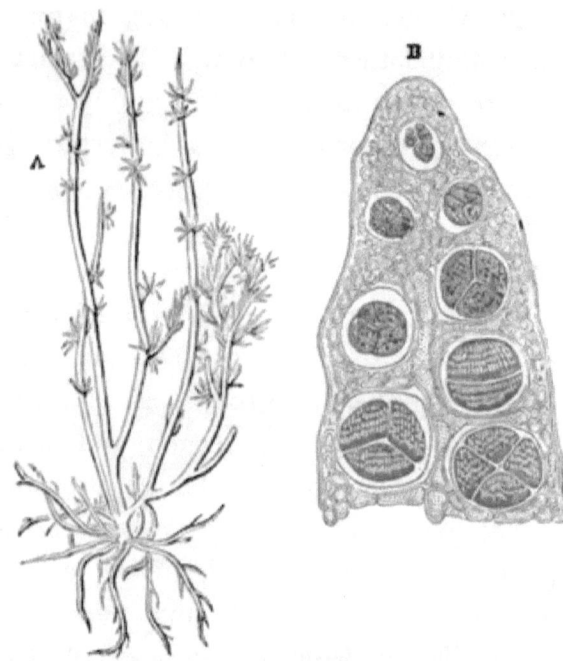

FIG. 36.—Arrangement of tretraspores, in *Carpocaulon mediterraneum:* A. Entire plant. B. Longitudinal section of branch. (N. B. Where only three tretraspores are seen it is merely because the fourth did not happen to be so placed as to be seen at the same view.)

It is in the form of globules, varying from the size of a pin's head to that of a hazel nut ; and as it grows freely, not being attached to any substance, it is readily driven by the wind from one place to another.

The thallus of Lichens may be of various sizes, forms, and colors.   (Fig. 37.)   Its fruit is called *Apothcceia*, and forms cups, or shields, of various forms, often colored bright red, yellow, gray, or black.   When these are divided by vertical sections they are found to contain a number of *asci*, or spore-cases, arranged vertically among filaments which are termed *paraphyses*. The fecundating

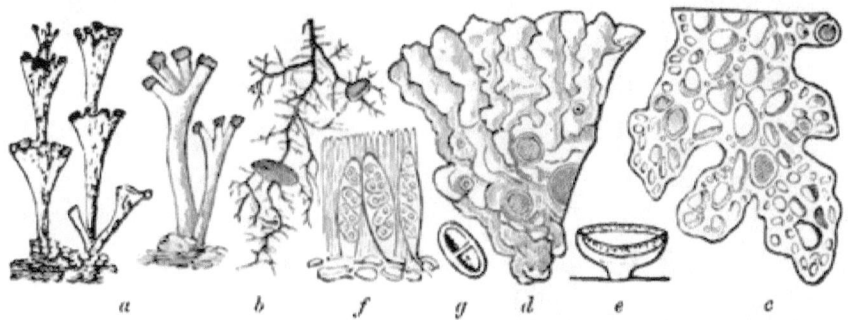

FIG. 37.—Lichens. *a*. Cladonia, with scarlet conceptacles. *b*. Usnea. *c*. Sticta. *d*. Parmelia. *e*. Vertical section of receptacle. *f*. Same highly magnified, with thecæ and paraphyses. *g*. Double spore.

apparatus is called the *Spermogonia*, and consists of small rounded or oblong organs, lodged in particular tubercles or immersed in the superficial layers of the thallus. The cellular filaments of the spermagonia give off minute oval bodies, called *spermatia*, which are analogous to the antherozoids of Algæ, but differ in being destitute of spontaneous motion.

7. The FUNGI form an extensive class of primitive organisms, generally ranked as plants, but which have so many peculiarities as to entitle them to be considered apart. We should not greatly err if we regard them as a third type of living things, differing both from animals and vegetables. They have no chlorophyll, as green vegetables have, and which enables them to break up carbonic acid. (Chap. VI, Sec. 2.) Light is not essential to the activity of Fungi, as it is to that of vegetables. They are incapable of assimilating inorganic food, but live upon organic substances. They are the agents of fermentation and of putrefaction, and their principal

office seems to be to break down and to restore to the
inorganic world the effete formed material of animal
and vegetable life.   Mushrooms, Puff-balls, Molds, and
the Rust of grain are examples of Fungi.

8. The simplest forms of Fungi resemble Protophytes,
except in the absence of chlorophyll, either green or red.
Recent investigations have indicated that those which
seem most simple are but imperfectly developed states
of other species.   The *Torula cerevisiæ*, or yeast-plant,
which is the cause of fermentation in solutions of sugar,

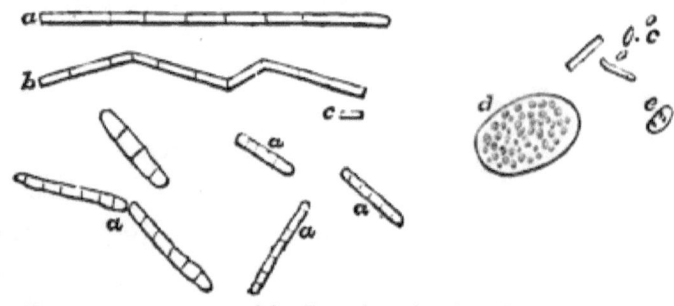

Fig. 38.—Appearances presented by Bacteria under the microscope. At *c*, are isolated Bacteria ; at *d*, they are arranged round a center ; while at *a* they appear in long
strings ; at *e* is observed a solitary torula.   All highly magnified.

and *Bacteria* (Fig. 38) of various forms, which cause putrefaction in animal substances, appear to be varieties, or
stages, in the development of some of the " molds," or
microscopic fungi, many of which are capable of polymorphism, or the assumption of many forms.   In some
kinds of fungi the bioplasm shows amœboid movements,
having a great resemblance to some of the lower forms
of animal life.

9. Fungi are cellular organisms of variable consistence.
They exhibit two well-defined structures, a *mycelium*, or
spawn, (Gr. *myces*, a fungus,) formed of filaments some-

times assuming a membraneous, a tubercular, or a pulpy form, and a *fruit*, or reproductive structure, which differs in different tribes. The essential reproductive organs are spores, called also *conidia*, usually four, or some multiple of four, attached to the cellular tissue, and supported on simple or branched filaments, (called *conidiophores*, or *basidia*,) or contained in sacs, (*thecæ, cystidia*, or *asci ;* all of which words, derived from the Greek, have similar meaning.)

10. Fungi have been divided into six orders, as follows :

1.) *Hymenomycetes*, (Gr. *hymen*, a membrane, and *myces*, a fungus.) Mycelium inconspicuous, bearing fleshy fruits which expand so as to expose the spore-bearing membrane to the air. Mushrooms are well-known examples.

2.) *Gasteromycetes*, (Gr. *gaster*, belly.) Fructifying surface inclosed, as in Puff-balls.

3.) *Coniomycetes*, (Gr. *konis*, powder.) The spawn or vegetative part is reduced to a minimum, and the abundant spores form a dusty or sometimes a gelatinous mass. The rust and bunt of wheat and other grains are instances.

4.) *Hyphomycetes*, (*hyphao*, to weave.) The vegetative part consists mostly of loose threads, as the naked seed Molds.

5.) *Ascomycetes*, (*askos*, a bag.) The sacs, or asci, containing the sporidia are either packed into an exposed hymenium, or line the interior of the fruit-bearing cysts, as Truffles, etc.

6.) *Physomycetes*, (*physa*, a bladder.) Mycelium filamentous, bearing sacs, containing minute porules, as the common Bread-mold.

11. The difficulty of determining what forms are to be regarded as species, and what as mere varieties, finds many illustrations in the class of Fungi. We know but little of the influence of external conditions in modifying forms, and the forms of fungi are so exceedingly unstable that the best observers are often at a loss. Yet this variability is only one of degree, since all living beings are more or less subject to modifications of form by external influences. It is this variability which has rendered the Darwinian hypothesis of evolution by "the survival of the fittest" so plausible a theory. But notwithstanding this capability of modification, there is still a certain fundamental and specific type for each assemblage of forms, and the amount of variability is strictly limited.

12. Many diseases of plants and animals are associated with the growth of Fungi. The "mildew" (*Puccinia*) and "rust" (*Uredo*) of wheat, etc., the potato blight, (*Peronospora*,) the disease in Silk-worms called Muscardine, (*Botrytis*,) the false membrane in diphtheria, the white patches in aphthæ, or thrush, and many skin affections, afford examples. Pyæmia is supposed to result from *bacteria* in the blood, and many epidemic diseases have been ascribed to similar origin. The prevalence of atmospheric changes, however, and variations in external conditions, as light, heat, moisture, etc., have much to do in predisposing both animal and vegetable tissues to disease, and in producing epidemics. Since the office of Fungi is to remove decaying or effete organic matter, we must discriminate between those diseased conditions which provide a habitat for fungi and the effects produced by the fungi themselves

13. The excessively minute and almost vapor-like sporules of fungi float about in the atmosphere in countless numbers, only waiting for a fitting soil in which to grow. As long as there is no refuse matter to be removed these scavengers are unemployed, but the smallest quantity of decaying animal or vegetable matter left exposed becomes covered with spores, which develop with astonishing rapidity. A scanty number of spores, only to be detected by careful research, will in a few days, and sometimes in a single night, give birth to myriads, to repress or remove the nuisances referred to. When the offal diminishes fewer of the spores find soil on which to germinate, and when all is consumed the active legions return to their latent or undeveloped state. Like Milton's spirits—

> So thick the aëry crowd
> Swarmed and were straitened; till, the signal given,
> Behold a wonder; they but now who seemed
> In bigness to surpass earth's giant sons,
> Now less than smallest dwarfs.

14. In the chapter on the Protophyte type of vegetable life, we considered the bioplasm, or living matter, differing in each kind, yet agreeing in one particular, namely, that each cell exhibits a repetition of the form and power of the parent-cell. In the Thallogens we find another idea predominating, or rather two leading ideas, the co-ordination of many cells in the structure of one individual, and the differentiation of cells in form and function, analogous to the division of labor in human society. Respecting the first, Joseph Cook has well remarked, in his axiomatic style: "Living tissues are co-

ordinated according to definite plans. As every change
must have an adequate cause, we are compelled to infer
the existence of a co-ordinating force behind the action
of the bioplasts in each organism. That force is the
cause of form in organisms. It has as many types as
there are types of organisms, vegetable and animal." *
The same subtle power which co-ordinates, also differen-
tiates the cells. This power resides in the original germ,
before the organization of the individual form, and is
what we have already defined as Life, and is not explica-
ble by physical causes. "Collocation of parts in an
organism is precisely what materialism has never yet
explained." †

* " Heredity," by Joseph Cook, p. 46.   † Cook's " Biology."

CHAPTER VIII.

ACROGENS.

Flower in the crannied wall,
I pluck you out of the crannies,
Hold you here in my hand,
Little flower, root and all,
And if I could understand
What you are, roots and all, and all in all,
I should know what God and man is.
—TENNYSON.

1. IN the type of Acrogens the instinct of development, or evolution of cells, is seen only at the summit or apex of the stem. Cells in other parts of the plant may enlarge, but do not multiply themselves. These plants generally have distinct stems and leaves, with stomata, (Chap. IV, Sec. 11,) a certain amount of vascular tissue, (Chap. IV, Sec. 10,) and *thecæ*, or cases containing spores. The Stoneworts, (*Characæ*,) the Liverworts, (*Hepaticæ*,) the Horsetails, (*Equisetaceæ*,) the Ferns, (*Filices*,) and the Mosses, (*Musci*,) are the principal families of Acrogens.

2. The Stoneworts (CHARACEÆ) have generally been regarded among Algæ, or water-weeds. But they differ greatly from Algæ in having a distinct axis of growth, and appendages. The axis may be a simple tube, (Genus *Nitella*,) or a tube with a cortical layer of smaller tubes surrounding it, (Genus *Chara*.) They are found in ponds and rivers, in tangled masses of a dull green color. Each plant is hardly thicker than a stout needle,

10

but may be three or four feet long. It has rootlets springing from the axis, by which it is fixed in the muddy bottom of the stream, etc., but the main source of its nutriment is the water in which it lives. It possesses chlorophyll, and in consequence decomposes carbonic acid under sunlight, retaining the carbon to form part of its own substance, and giving off the oxygen. The branchlets (or *leaves*, as they are called) are grouped in whorls, or spring from the same height in the stem.

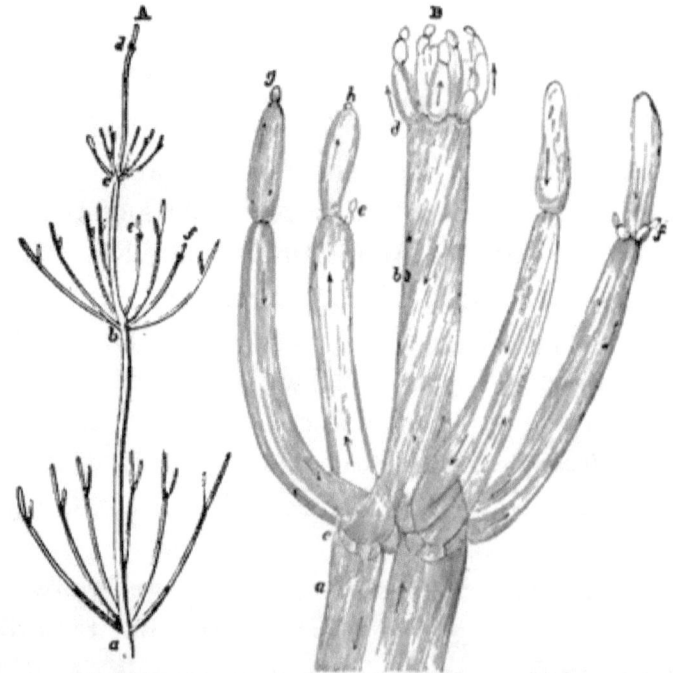

FIG. 39.—*Nitella flexilis:* A. Stem and branches of the natural size. *a. b. c. d.* Four verticils of branches issuing from the stem. *e. f.* Subdivision of the branches. B. Portion of the stem and branches enlarged. *a. b.* Joints of stem. *c. d.* Verticils. *e. f.* New cells sprouting from the sides of the branches. *g. h.* New cells sprouting at the extremities of the branches.

and at regular intervals. (Fig. 39.) The main stem is called the axis, and a branch, when it exists, is a secondary stem. The appendages are the leaves, branches, rootlets, and reproductive organs. The points on the axis, or stem, from which the appendages spring, are called *nodes*, the intervening parts being the *internodes*. In Chara the internodes have a spiral striation. Growth takes place at the apex by the development of new nodes and internodes. Each internode is formed by the growth and elongation of one cell.

The terminal bud is formed by a single cell, which divides by fission into two, one of which forms the internode, while the other subdivides into lateral cells, which, by division, produce the appendages. In the latter, after a certain stage, the terminal cell is incapable of further division, but in the stem the process continues indefinitely. (Fig. 40.)

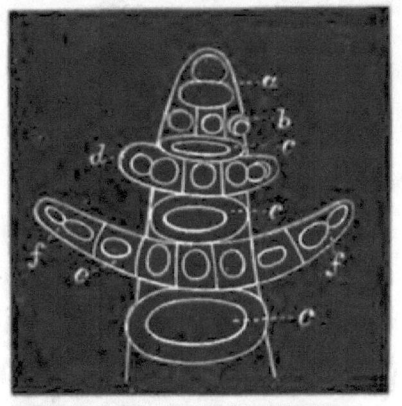

FIG. 40.—Growing Point of Chara. *a.* Terminal cell dividing. *b.* Cells forming youngest node, and which by their fission will give rise to a whorl of appendages. *c. c.* Internodal cells. *d.* Incipient appendages. *e.* Same farther advanced. *f. f.* Terminal cell dividing.

The reproductive organs in these plants are of two kinds, oval *sporangia*, or spore-fruits, and *antheridia*. The latter are smaller than the sporangia, and globular, and contain filaments whose cells are changed into little ciliated bodies called *antherozoids*. (Fig. 41.)

The growing spore from the sporangium gives off two filaments resembling the hyphæ in fungi, one of which serves as a temporary root, and a cell in the other produces a group of lateral projections from which the young

Chara springs. This temporary structure is termed the *pro-embryo.*

In *Chara vulgaris* the circulation, or movement, of

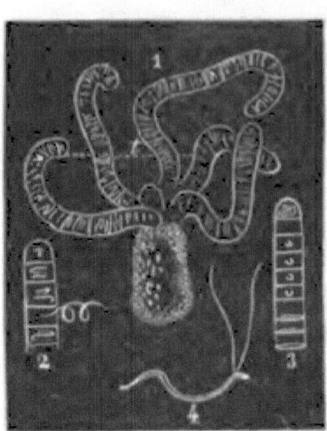

bioplasm in vegetables was first discovered. (Chap. II, Sec. 7, and Chap. IV, Sec. 6.)

3. The Liverworts, or HEPA-TICÆ, form a class or group of plants generally considered intermediate between Lichens and Mosses. They are furnished with leaves, or lobed fronds, with rootlets on the under surface, which send up stalks carrying either round, shield-like disks, bearing *an-theridia,* or radiating bodies,

FIG. 41.—Portion of Antheridium of Chara. 1. Several jointed filaments attached to a vesicle. 2. End of one of the tubes, a spiral thread escaping. 3. A tube nearly empty. 4. An antherozoid with its cilia.

bearing at first *archegonia,* or female organs, and afterward *sporangia,* or spore-cases. (Fig. 42.)

The arrangement of the stomata, or breathing pores, in these humble plants is far more complex than we find it in others. The leaves of all the higher plants have cavities, or air-spaces, communicating with the external world

FIG. 42.—Frond of *Marchantia polymorpha,* with gemmiparous conceptacles, and lobed receptacles bearing pistillidia.

by openings, or pores, which are guarded by elastic cells; but in *Marchantia polymorpha* the green surface of the frond is seen by a low magnifying power to be divided into diamond-shaped spaces, containing an opening in each. On making a thin section, as in Fig. 43,

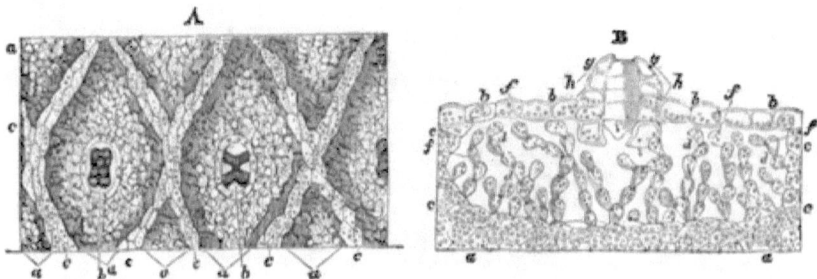

FIG. 43.—A. Portion of frond of *Marchantia polymorpha* seen from above. *a. a.* Lozenge-shaped divisions. *b. b.* Stomata seen in the center of the lozenges. *c. c.* Greenish bands separating the lozenges. B. Vertical section of the frond, showing *a. a.* the dense layer of cellular tissue forming the floor of the cavity *d. d.* *b. b.* Cuticular layer, forming its roof. *c. c.* Its walls. *f. f.* Loose cells in its interior. *g.* Stoma divided perpendicularly. *h.* Rings of cells forming its wall. *i.* Cells forming the obturator ring.

each of these stomata will be seen to form a sort of shaft, or chimney, of four or five rings, or courses, of cells. the lowest ring regulating the aperture into the leaf-grottoes below.

The spores of Marchantia are attached to *elaters*, or spirally-coiled elastic fibers, whose extension scatters the spores.

3. The EQUISETACEÆ, or Horsetails, are found in most parts of the world, save Australia and New Zealand. They generally grow in wet places, sending up shoots from a creeping stem, or *rhizome*. The cuticle is remarkable for the great quantity of silica contained in it. The particles of this mineral, each having a double axis of refraction, are arranged in rows parallel to the axis, and are beautiful objects under the microscope, with polarized light. The abundance of silica has led

10*

to some of these plants being used as natural files for polishing various articles.

The shoots are jointed, each articulation having a

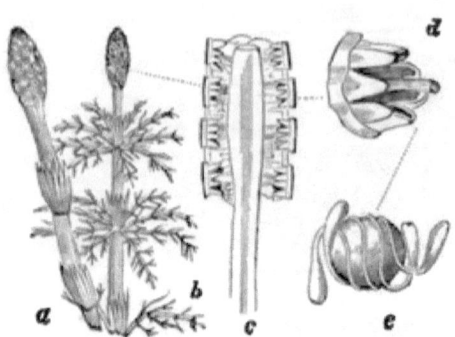

toothed membrane-ous sheath, and having whorls of branches and branchlets. The fructification is in the form of terminal cones, with scales bearing spore-cases, and opening by a longitudinal fissure. Each of the spores

FIG. 44.—*a.* Equisetum arvense. *b.* Equisetum sylvaticum. *c.* Section of the spike. *d.* A sporange. *e.* A spore with its elaters coiled.

has a pair of spiral filaments, with clubbed ends, and attached by their center, so as to look like four stamens. (Fig. 44.)

4. FERNS in tropical countries are sometimes rivals of the most beautiful Palms, having trunks varying from two or three to sixty or eighty feet in height, formed of the consolidated bases of the fronds. In these Tree-ferns the fronds are either repeated in whorls, or they form a tuft at the summit, constituting in the latter case a collection of whorls with suppressed internodes. In the ordinary ferns, or brakes, of temperate climes, the stem is an underground one, or *rhizome*, and the disposition of the fronds is seldom observed.

The epidermis of the stem is of a brownish hue, and the general cellular structure, or *parenchyma*, consists of many-sided nucleated cells, containing chlorophyll and starch granules. There are also vessels (annular, spiral,

and scalariform, or ladder-like) and fibrous or woody
tissue, forming together the *sclerenchyma*, or harder tissues. (Chap. IV, Secs. 6 to 10.)

In Fig. 45 the acrogenous growth of a fern is illustrated, together with the metamorphosis of the terminal

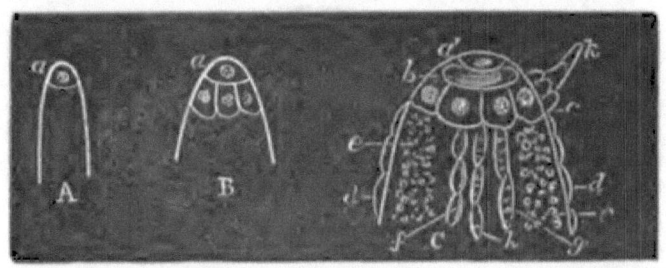

FIG. 45.—Diagram, showing the mode of growth in the stem of a Fern. A. B. C.
Stems of ferns showing successive stages of growth. *a. a. a.* Terminal cells, the latter
just after being produced by division. *b.* A cell which will give rise to an internode.
c. Shows a ring or cluster of cells giving rise to a node. *d.* Epidermal cells. *e.* Parenchyma.
f. Sclerenchyma. *g.* Scalariform vessels. *h.* Spiral vessels. *k.* An appendage, originating
at the node. *d. e. f. g.* and *h.* all arise from the multiplication and metamorphosis of the
"growing" cells.

cell into the various tissues. In flowering plants the
terminal cell of the leaf-bud becomes barren, and the
enlargement of the leaf depends on the multiplication and growth of cells nearer the base; but in the
fern the leaf-bud grows as the stem does, so that the
peduncle is first formed, then the embryo leaf, then the
pinnules, etc.

Underneath the frond of a fern we may sometimes see
little brown patches. Each patch is a *sorus*, (sometimes
covered by a membrane called an *indusium*,) and the
little brown bodies constituting it are *sporangia*, or
spore-cases, which have been developed from epidermal
cells. An elastic ring (*annulus*) surrounds each sporangium, and assists in opening it. The growth of these

minute spores may be observed by sowing them on a saucerful of fine mold, covering with a bell-glass or tumbler, and keeping it moist, warm, and shaded. A green film will spread over the soil, which can be taken up, from time to time, on the point of a knife for microscopic examination. The little spore swells and bursts, and throws out a rootlet which gets its nourishment from the soil. Then a number of delicate cells are formed from the mother-cell in the spore, making a little green scale, (the *prothallium*,) which throws out rootlets on its under side. This prothallium produces two kinds of cells, one set called *antheridia*, which contain spiral filaments which escape to enter the others, called *archegonia*, or germ-cells, from which the future fern is reproduced. (Fig. 46.)

The fossil remains of Ferns in the various strata of the earth's crust are very numerous, especially in the Coal measures. These deposits exhibit the remains of many species now extinct. Immense tree-ferns and gigantic Lycopodiaceæ (Club-mosses) flourished in an atmosphere charged with moisture and carbonic acid gas, which, by plant assimilation and liberation of oxygen, is thought to have been purified and prepared for the use of successive tribes of animals and of man.

5. MOSSES are minute and lowly plants, but they are by no means insignificant. There are about ten thousand species, some of which are not over a hundredth part of an inch in height, while others are several inches high. Mosses have a distinct axis of growth, and the delicate leaves are arranged with great regularity. The stem shows some indication of the separation of a cortical, or bark-like portion, from the medullary, or pith-

like, by the intervention of a circle of bundles of elongated cells, from which prolongations pass into the leaves,
so as to afford them a sort of midrib.

FIG. 46.—Ferns and their parts. *a*. Fronds and root-stalk. *b*. Frond, showing the
spore-cases. *c*. Exterior and interior of seed-vessel. *d*. Fronds, gradually unfolding. *e*. A
Theca, or spore-case, before opening. *f*. A Theca, or spore-case, discharging its spores.
g. Prothallus of its natural size. *h*. Lower surface of prothallus, much enlarged, showing
the organs whose reciprocal action determines the development of the fern. *i*. Various
forms of one of these organs when in movement. *j*. Inclosed vesicle, in which the development of the fern commences.

The root-fibers are long tubular cells, quite transparent, within which the circulation of the bioplasm may
be seen. Dr. Hicks has observed portions of the inclosed
bioplasm detached, and having amœboid motions.

The stems of Mosses usually terminate in filaments, or foot-stalks, supporting an urn-shaped vessel closed by a lid, (*operculum,*) which is covered by a cap, or hood, (*calyptra.*)   Under the operculum, the edge of the urn is a beautiful toothed fringe, (the *peristome,*) and within the

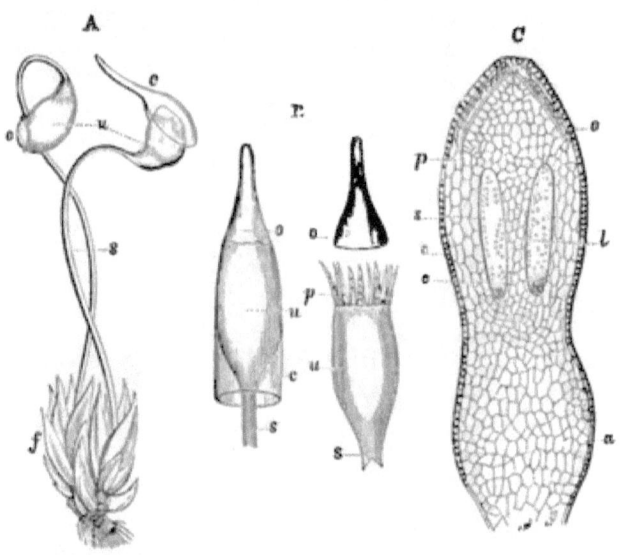

FIG. 47.--Structure of Mosses. A. Plant of *Funaria hygrometrica*, showing, *f.* the leaves ; *u.* the urns supported upon the setæ, or footstalks, *s*, closed by the operculum, *o*, and covered by the calyptra, *c.* B. Urns of *Encalyptra vulgaris*, one of them closed and covered with the calyptra, the other open. *u. u.* The urns. *o. o.* The opercula. *c.* Calyptra. *p.* Peristome. *s. s.* Setæ. C. Longitudinal section of very young urn of *Splachnum.* *a.* Solid tissue forming the lower part of the capsule. *c.* Columella. *l.* Loculus, or space around it for the development of the spores. *e.* Epidermic layer of cells, thickened at the top to form the operculum, *o.* *p.* Two intermediate layers, from which the peristome will be formed. *s.* Inner layer of cells forming the wall of the loculus.

urn, or spore-capsule, (*sporangium,*) are double-coated spores. (Fig. 47.)

In developing into new plants, the outer coat of the spore bursts and the inner wall protrudes. New cells grow from the extremity, forming a filament, whose cells at certain points multiply by subdivision, so as to form rounded clusters, like the prothallus of Ferns, or the pro-embryo of Chara, from each of which an independent plant may arise.

The urn, or capsule, is not the real fructification of a moss, but its product, since Mosses, like Liverworts, etc., possess both antheridia and pistillidia, although they are often inconspicuous. They are found either together, or on different parts of the same plant, or on different plants. They are usually situated at the bases of the leaves, close to the axis. The *antheridia* are globular or elongated capsules containing sperm-cells, each of which produces a moving antherozoid, which escapes at the summit of the capsule. Hair-like filaments (*paraphyses*) around the antheridia are considered to be sterile or undeveloped antheridia. The *archegonia* are like those of the Hepaticeæ, and when fertilized the embryo-cell develops by cell-division into a conical body elevated upon a stalk. This tears the walls of the flask-shaped archegonium by a circular fissure, carrying the upper part as the *calyptra*, or hood of the "urn," while the lower part remains as a sort of collar around the stalk.

6. The characteristics of the type of Acrogens are the growth of cells at the summit only; the appearance of a distinct cortical portion, or epiderm, and of vascular and fibrous tissue; and a sort of alternation of generations in the provision of a prothallus, so that plants of this type may not improperly be designated Prothallus plants, as the higher types are sometimes known as Monocotyledonous, or Dicotyledonous plants.

7. As an illustration of the reflections natural to a well-regulated mind from the study of natural objects,

even of minute and apparently insignificant Acrogens,
an incident in the life of Mungo Park is appropriate.
This enterprising traveler, during one of his journeys
into the interior of Africa, was robbed and stripped by
banditti. When the robbers had left him, he says: "I
sat for some time looking around me with amazement
and terror. Whichever way I turned, nothing appeared
but danger and difficulty. I found myself in the midst
of a vast wilderness, in the depth of the rainy season,
naked and *alone*, surrounded by savage animals, and men
still more savage. I was five hundred miles from any
European settlement. All these circumstances crowded
at once upon my recollection, and I confess that my
spirits began to fail me. I considered my fate as certain, and that I had no alternative but to lie down and
perish. The influence of religion, however, aided and
supported me. I reflected that no human prudence or
foresight could possibly have averted my present sufferings. I was indeed a stranger in a strange land, yet I
was still under the protecting care of that Providence
who has condescended to call himself the stranger's
friend. At that moment, painful as my reflections were,
the extraordinary beauty of a small moss irresistibly
caught my eye, (I mention this to show from what
trifling circumstances the mind will sometimes derive
consolation,) and, though the whole plant was not larger
than the top of one of my fingers, I could not contem-
plate the delicate conformation of its root, leaves, and
fruit without admiration. Can that Being (thought I)
who planted, watered, and brought to perfection, in this
obscure part of the world, a thing which appears of so

small importance, look with unconcern upon the situa-
tion and sufferings of creatures formed after his own
image? Surely not. Reflections like these would not
allow me to despair. I started up; and, disregarding
both hunger and fatigue, traveled forward, assured that
relief was at hand, and was not disappointed." Such
views of the universe and of its Creator infuse strength
into the human soul, and give a vigor to human character which is impossible otherwise. For the good-order-
ing of human life they are infinitely above all Monistic
speculations and theories of Evolution, which belittle
or lose sight of the individual in a romantic sentiment
respecting primordial and progressive development of
all.

11

CHAPTER IX.

ENDOGENS.

What time this world's great workmaister did cast
To make all things such as we now behold,
It seems that He before His eyes had plast
A goodly patterne, to whose perfect mould
He fashioned them as comely as he could,
That now so fair and seemly they appear
As naught may be amended anywhere.

—SPENSER.

1. ENDOGENOUS plants have no separable bark, nor distinct concentric circles in the stem. Their fibro-vascular bundles, consisting of spiral and porous vessels and woody fibers, descend from the leaves downward, converging at first toward the center, but afterward diverging outward until they reach the roots, or attach themselves to the hardened tissue of the outer or cortical layer, corresponding to the bark in Exogens, but harder than the rest of the stem, and inseparable from it. It used to be thought that the woody portion was added to the center, and pushed the first-formed fibers toward the circumference; hence the term Endogenous, (*endon*, within, and *gennao*, to produce.) In strict scientific accuracy the term only applies to the fibers at the early part of their course, since in the latter part they become blended in the cortical layer, forming a tough net-work. The center of the stem when young is filled with cells which sometimes disappear, except at the nodes, leaving the stem hollow, as in Grasses.

The embryo of Endogens has but a single *cotyledon,* or seed-lobe, on which account they are often termed Monocotyledons. Acrogens have no seed-lobe, but cellular spores, and are called acotyledons, (from the Greek *a,* privative, and *kotyledon,* something hollow,) while Exogens have two seed-lobes, and are dicotyledons.

The veins in the leaves of Endogens are generally parallel, or straight, (Fig. 48,) and do not form a net-work, and the parts of the flower are arranged in sets of threes, or of some multiple of three.

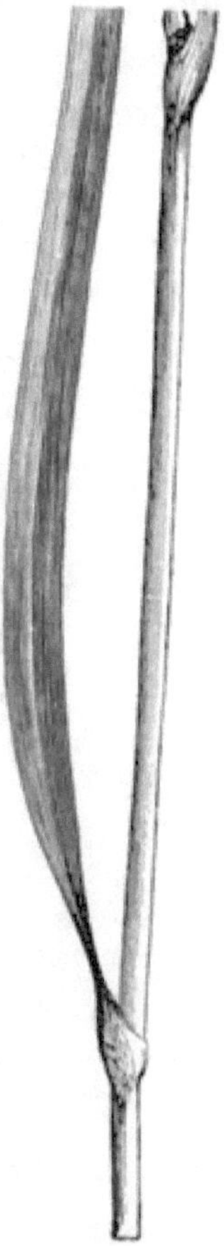

2. Exceptions to the parallel venation of leaves in Endogens have been placed by Lindley in a class by themselves—the *Dictyogenæ,* (from *dictyon,* a net, and *gennao,* to produce ;) in allusion to the reticulation of the leaves. They comprise the Yam tribe, (*Dioscoreaceæ,*) the Sarsaparilla family, (*Smilaceæ,*) and the Trillium family, (*Trilliaceæ.*) The other classes or sub-classes are, 1. *Petaloideæ,* or *Floridæ,* in which the flowers consist either of a colored perianth (a floral envelope) or of scales arranged in a whorl. 2. *Glumiferæ,* in which the flowers have imbricated bracts or scales, called *glumes.* This includes the two orders of grasses and sedges. The *Petaloideæ* are divided into three sections: 1. *Epigynæ,* having perfect flowers, and a superior perianth, (ovary inferior,) as Orchids, Gingers, Irids, Amarylids, etc. ; 2. *Hypogynæ,* having perfect flowers

FIG. 48.—Endogenous Leaf, showing parallel venation.

and an inferior perianth, (ovary superior,) as Lilies, Rushes, and Palms; 3. *Incompletæ*, with imperfect flowers, without a proper whorled perianth, as Screw-pines and Arums.

3. GRAMINEÆ, the Grass Family, is one of the most important orders in the vegetable kingdom, whether we regard it as supplying food for man or herbage for animals. Grasses are found in all quarters of the globe, and are said to form about 1-22d part of known plants. There are about 4,000 species, and their structure is the most simple of the higher forms of vegetation. Their stems form protecting sheaths to the growing shoots, and have alternate leaves. Their flowers, or glumes, present many varieties, producing the distinctive characters of families or tribes, and genera. Among the grasses are the nutritious cereal grains, as Wheat, (*Triticum*,) Oats, (*Avena*,) Barley, (*Hordeum*,) Rye, (*Secale*,) Rice, (*Oryza*,) Maize or Indian Corn, (*Zea*,) etc. Here, also, are found various pasture grasses, as Rye-grass, (*Lolium*,) Timothy-grass, (*Phleum*,) Meadow-grass, (*Poa*,) etc. (Fig. 49.)

FIG. 49.—Wheat, Barley, Meadow-grass.

The cereal grains have been so generally distributed by man that all traces of their native country are lost. They seem to be examples of permanent varieties or races preserved by cultivation.

The grains, or seeds, of many kinds are used for food, since they contain a large amount of starch and gluten.

Sugar is also obtained from many grasses, as the Sugar-cane, (*Saccharum officinarum,*) Sweet Sorghum, (*Sorghum saccharatum,*) etc.

Grasses contain a large quantity of silicious matter in the epidermis of their stalks, which sometimes accumulates in the joints, as the Tabasheer in the joints of Bamboo, (*Bambusa.*) This latter is a tree-like grass, sometimes growing fifty or sixty feet high. It is applied to an almost endless variety of purposes. The Chinese use it, in one way or other, for nearly every thing they require. The sails of their ships, as well as their masts and rigging, and articles of furniture, as mats, screens, chairs, tables, bedsteads, and bedding, are all made out of the Bamboo, which is cultivated with great care.

The stems of some grasses run under-ground, and are useful in consolidating the sand of the sea-shore. This property renders some grasses (as *Triticum repeus*) difficult to exterminate.

4. SEDGES, (*Cyperaceæ,*) are grass-like herbs, with angular stems and narrow, tapering leaves wrapping round the stem, but without the slitting sheath. Their flowers are borne on bracts, or scales, united in an imbricated manner so as to form a spike. In Lapland they equal the grasses in number, but the proportion decreases toward the equator. Few plants of this family are attractive to the eye, but many of them are useful. The creeping stems of *Carex arenaria* bind the shifting sands on the shores of Brittany and Holland into a wind-defying mass. The *Papyrus antiquorum* of the Nile (the Bulrush of Scripture) belongs to this family. It for-

11*

merly furnished the world with paper, besides being
used for making boats, ropes, etc.   (Fig. 50.)

FIG. 50.—The Papyrus of the Nile.   FIG. 51.—Cocoa-nut Tree, (*Cocos nucifera*,)
and Plantain, (*Musa Paradisiaca*.)

5. Linnæus, the great Botanist, called Grasses the
plebeians, and PALMS the princes, of the vegetable

world. The latter are for the most part trees of gigantic growth, often reaching dimensions unknown among other plants. They are used for supplying food and for forming habitations. The fruit of some is edible, while that of others is hard. Many supply oil, wax, starchy matter, and sugar, which is fermented to form an intoxicating beverage. Their fibers make ropes, and the reticulum about their leaves is sometimes manufactured into brushes.

The Date palm, (*Phœnix dactylifera,*) which supplies food to so many of the inhabitants of Arabia and Africa, is considered to be the Palm of the Bible. The Cocoanut palm (*Cocos nucifera*) is one of the most useful, supplying the South Sea Islander with food, clothing, houses, utensils, ropes, oil, sugar, wine, and Palm cabbage from the terminal bud, etc. (Fig. 51.)

Sago and other starchy matter is obtained by bruising and washing the cellular tissue of many Palms, especially *Sagus Rumphii, S. lævis* and *S. genuina.*

6. The BANANA family (*Musaceæ*) contains plants which furnish a large supply of nutritious fruit, while their leaves afford valuable fibers. The best known species are *Musa paradisiaca,* or Plantain, and *M. sapientum,* or Banana, (Fig. 52;) the former a denizen of the Old World, and the latter of the New. The specific name of the first originated in a notion of some of the old botanists that it was the forbidden fruit of Eden. A quaint writer remarks that it is not likely that a plant so useful should have been the forbidden fruit. The Banana supplies the inhabitants of the tropical islands with wholesome and abundant food, pleasant drink, valuable

medicine, materials for clothing, baskets, mats, and with almost all other necessaries of their simple life. *Musa*

FIG. 52.—Banana, (*Musa sapientum.*)

textilis supplies a flax-like fiber, from which some of the finest Indian muslins are made.

7. The ARROW-ROOT family, (*Marantaceæ,*) the PINE-APPLE family, (*Bromeliaceæ,*) and the GINGER family, (*Zingiberaceæ,*) contain many useful species of Endogens. Here also are classed many of the showy flowers of our gardens and hot-houses. The IRIS family, (*Iridaceæ,*) containing the Iris, Gladiolus, and Crocus, etc.; the singular aquatic plants of the Hydrocharidaceæ, as Hydrocharis and Valisneria; and the AMARYLLIS family, (*Amaryllidaceæ,*) embracing the Daffodil, Amaryllis, and

Agave, are of this class. The last-named plant is sometimes called the American Aloe, (*Agave Americana*,)
and, according to gardening fable, only blooms once in
a hundred years, hence called Century plant. Its large,
hard, spinous leaves grow slowly for years, when suddenly, in the course of a single season, a stem shoots up forty
or fifty feet in height, bearing a crest of flowers. In
Peru and Mexico an intoxicating beverage called *pulque*
is made from the sap.

8. The ORCHID family (*Orchidaceæ*) exhibit the great-

FIG. 53.— Orchids, (*Orchidaceæ*.)

est variety of forms and brilliancy of color among all the
vegetable tribes. The flowers often resemble insects, as

butterflies, moths, flies, and spiders; or birds, as doves and eagles; or reptiles, as snakes, lizards, and frogs. (Fig. 53.) Their spots and colors give sometimes the appearance of leopard or tiger skins. These resemblances are indicated in their generic and specific names. Some parts of the petals of these flowers display peculiar irritability, for the purpose of scattering the fertilizing pollen. As the visits of insects are often subsidiary to the fertilization of Orchids, they have attracted much attention from naturalists. They abound chiefly in moist tropical climates.

9. The LILY family (*Liliaceæ*) includes many showy garden flowers, as Tulips, Lilies, Dogtooth-violets, and Tuberoses. (Fig. 54.) It is divided into several tribes, as the Onion, or Squill tribe, the Asphodel tribe, the Lily-of-the-valley tribe, the Aloes tribe, and the Asparagus

FIG. 54.—The White Lily. (*Lilium album.*)

tribe. In the latter tribe are placed the Dragon-trees, the most gigantic of the order. There is one in the Island of Teneriffe which is described as seventy feet high and forty-six feet in circumference at the base. The flowers are small. From some species of Dragon-tree the Sandwich Islanders prepare an intoxicating liquor called *ava*.

The inspissated juice of several species of Aloe is used in medicine as a cathartic, and the bulb of the Squill is imported from the coasts of the Mediterranean, and is valued for its diuretic, expectorant, and other properties.

A species of Onion called Camass is used by the Indians of Oregon as food.

Textile fibers are procured from New Zealand flax (*Phormium*) and from the *Yucca*, or Adam's needle.

10. The SCREW-PINE family (*Pandanus*) contains several species which exhibit a semblance of instinct in the development of aerial roots at different distances on the stem, by which their life is prolonged. Their leaves are arranged in a spiral, hence the name, Screw-pine.

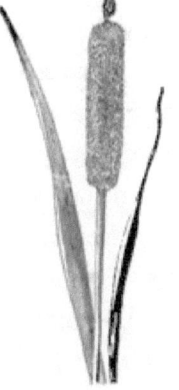

FIG. 55.—Bulrush.

11. The ARUM family contains the Cuckoo-pint tribe, the Bulrush tribe, (Fig. 55,) the Sweet-flag tribe, and the Duck-weed tribe. In the Duckweed (*Lemna*) we see at a casual glance nothing but a green scale floating on the water, which is in reality a compound of both root and stem. A careful observation in summer may lead to the discovery of minute straw-colored anthers on the edges of the plants,

and near these a narrow slit, which, on being enlarged, will show the simple flower, like a membraneous bag, and containing two stamens and one ovary, with its style and simple stigma.

11. Protophytes, Thallogens, and Acrogens have been classed together in the artificial system of Linnæus as *Cryptogamia*, (from *cryptus*, hidden, and *gamos*, nuptials,) in allusion to the inconspicuous character of their reproductive organs ; while Endogens and Exogens are called *Phanerogamia*, (*phaneros*, visible, and *gamos*, nuptials,) since they have perceptible reproductive organs formed of *stamens* and *pistils*. To these essential parts we frequently find two envelopes added, the *calyx* and *corolla*. These parts make up the flower, and the Phanerogamia are not infrequently known as flowering plants.

The flower consists of whorled leaves placed on an axis, the internodes of which are not developed. There are usually four of these whorls. The outer whorl is the calyx, the next the corolla, the third the stamens, and the innermost the pistil. In Exogens the calyx is usually green and the corolla colored, but in Endogens both frequently display rich coloring, and are apt to be confounded, so that the term *perianth* is usually applied to the flowers of Endogens, whether colored or otherwise, (*peri*, around ; *anthos*, flower.)

The parts of the calyx, when separate, are called *sepals*, and the leaves of the corolla *petals*. Stamens have two parts, the *filament*, or stalk, and the *anther*, or broader portion, corresponding to the folded blade of the leaf, and containing fertilizing grains called *pollen*. The pistil is also made up of two parts, the *ovary*, con-

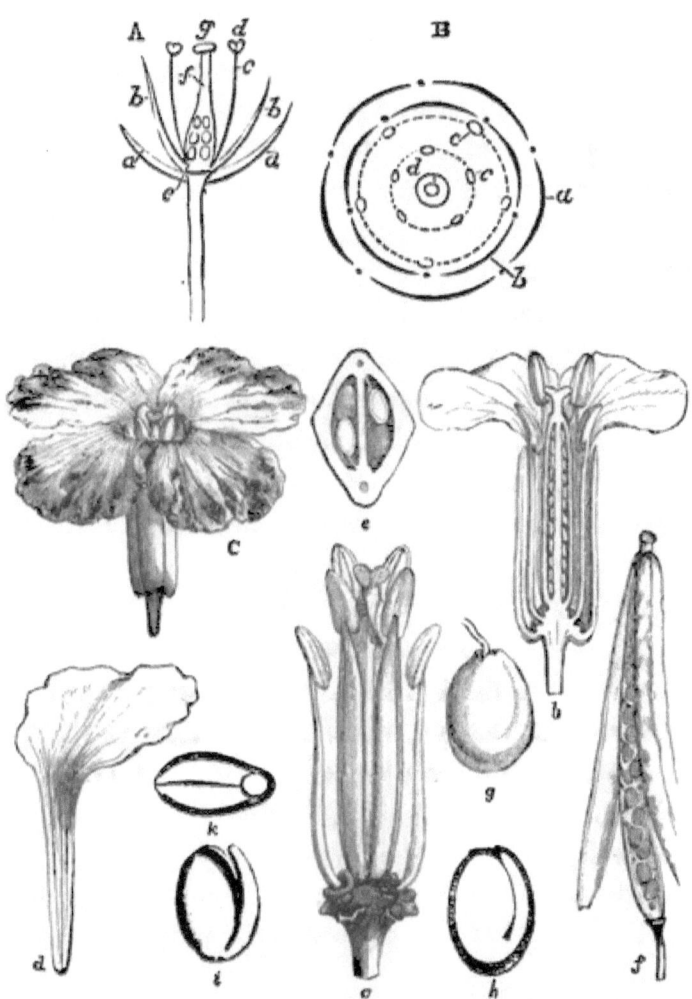

FIG. 56.—A. Sectional view of the flower, showing the vertical disposition of the whorls. *a*. Sepal of calyx. *b*. Petal of corolla. *c*. Filament of stamen. *d*. Anther of stamen. *e*. Ovary of pistil. *f*. Style of pistil. *g*. Stigma of pistil. B. Plan of the typical flower of an exogenous plant, showing the horizontal disposition of its parts. *a*. Sepal. *b*. Petal. *c.c*. Stamens in two distinct whorls. *d*. Carpel or ovary, inclosing an ovule, attached by its funiculus. C. Various parts of the clove. *a*. Flower of the clove or pink. *b*. Vertical and middle section of the flower. *c*. Flower reduced to its male and female portions; the stamens are six in number—four large, (in pairs,) and two small. *d*. One of the petals. *e*. Horizontal section of the ovary, or seed-vessel; showing the insertion of the ovules. *f*. Fruit at the moment of expansion. *g*. Seed, with its funiculus. *h*. Vertical section of the seed and its embryonic contents. *i*. The embryo alone. *k*. Horizontal section of the seed and its embryonic contents.

taining ovules or young seeds, and the *stigma*, a cellular secreting body for the reception of the pollen-grains. This is sometimes sessile, or resting on the ovary, and sometimes elevated on a stalk, or *style*. Like the other whorls, the pistil is made up of one or more modified leaves, named *carpels*. (Fig. 56.)

Some flowers have no stamens, and are called female flowers; others have no pistil, and are male flowers: but both stamens and pistils are always present, either on the same plant or on different plants. Some flowers have neither calyx, corolla, nor stamens; others, neither calyx, corolla, nor pistil. If they have no corolla they are *incomplete*, and if corolla and calyx are both absent they are *naked*.

The general axis of inflorescence is called *rachis;* the stalk supporting a flower or a cluster of flowers is a *peduncle*, and, if small branches are given off by it, they are called *pedicels*. Sometimes the floral axis is shortened, and is flat, convex, or concave, bearing numerous flowers, as in the Daisy. It is then called a *receptacle*.

Flowers are always the termination of an axis, branch, or bough, and the order governing their arrangement is a repetition of that which governs the ramification of the plant.

Bracts, or floral leaves, are leaves from which the floral axis, or the individual flowers, arise. Sometimes they are colored and may be mistaken for parts of the corolla, and at other times they are undeveloped. Bracts are generally deciduous, but occasionally persist, and even form part of the fruit, as in the cones of Firs and the fruit of the Pine-apple. In catkins (or imperfect unisex ual sessile flowers on a spike, as in the Willow or Hazel) the bracts are called *scales*. A whorl of bracts is an *involucre*. These are sometimes adherent, as in the cup of the Acorn. A sheathing bract inclosing one or more flowers is a *spathe*. This is common among Endogens, as in Calla, Arum, and the Palms. In Grasses the outer scales are considered as sterile bracts, and have received the name of *glumes*.

The various modes of inflorescence is a subject of profound study with botanists, but its details are too extensive for the design of the present work. As stated in Chap. IV, Sec. 11, the parts of the flower, as regards their development, structure, and arrangement, may all be referred to the leaf as a type. They begin like leaves in cellular projections, in which fibro-vascular tissue is ultimately formed; they are arranged in a more or less spiral manner, and they are often partially or entirely changed into leaves. These facts confirm Goethe's doctrine that all the parts of the flower are altered leaves.

12. In the type of Endogens we meet with a great variety of flowers, some perfectly organized, as the Lily, and others, as the Duckweed and Bulrushes, quite incomplete. Yet even in the more lowly forms we meet with abundant examples of the care of a beneficent Providence accomplishing intelligent designs by various ways, but all indicative of Divine wisdom. In the Branched Bur-reed (*Sparganium ramosum*) the branches bear yellow balls of staminate, or barren flowers, and green pistillate, or fertile florets. "What happens in this case," says Dr. Lindley, "occurs also in all instances in which the stamens are separated from the pistils in

different flowers on the same plant; we invariably find
that the stamens are placed on the uppermost parts of
the branch above the pistils, an arrangement which is
no doubt provided to facilitate the scattering of their
pollen upon the stigmas. If they were placed below
the pistils it would be much more difficult for the pollen to reach the stigma, and consequently, the great
end of the creation of the stamens would be almost
frustrated. We find, however, that every thing is foreseen and provided for by Providence, with the same
care in these little plants as in the most exalted and
perfect of the works of nature; and that even so apparently useless and insignificant a weed as the Bur-reed
contains the most convincing evidence of the worthless-
ness of the opinions of those who, denying the existence
of the Deity, would have the world believe that living
things are the mere result of a fortuitous concourse of
atoms, attracting and repelling each other with different
degrees of force." *   The recent elaborate observations
of Darwin, Sir J. Lubbock, and others, upon the fertilization of plants, as the Orchids, by the visits of insects,
although sought to be explained by the principle of
"unconscious natural selection," finds a more ready and
satisfactory explanation in the case of an ever-present
Providence, since if the colors, and honey, and structure
of the flowers are "all arranged with reference to the
visits of insects," † the structure and habits of insects
are equally adapted to the fertilization of the flowers.
From the design we infer a Designer.

* "Ladies' Botany," by Dr. J. Lindley.
† Lubbock's "Wild Flowers in Relation to Insects."

CHAPTER X.

EXOGENS.

In all places, then, and in all seasons,
   Flowers expand their light and soul-like wings,
Teaching us, by most persuasive reasons,
   How akin they are to human things.
And with childlike, credulous affection,
   We behold their tender buds expand,
Emblems of our own great resurrection,
   Emblems of the bright and better land.
             —LONGFELLOW.

1. THE term Exogen (from *exo*, outward, and *gennao*, to produce) is applied to those plants which produce woody and vascular layers toward the circumference. It is the largest class, or type, in the vegetable kingdom, including about 7,000 genera and 70,000 species of flowering plants.

External to the woody layers, and between them and the bark, is a layer of semifluid mucilaginous matter called *Cambium*. Its cells are exceedingly delicate. New cells are continually being added, on the inner side of the Cambium layer, to the thickness of the wood, and on the outer side of it, to the thickness of the bark, increasing the diameter of the axis of the plant.

At the apex of the stem, and at that of the root, the Cambium layer is continuous with the cells which retain the power of dividing in these localities.

The general appearance of the axis of an exogenous plant is that of a double cone; one cone representing

12*

the stem, the other the root; the growing part of both being bathed in the cambium fluid.

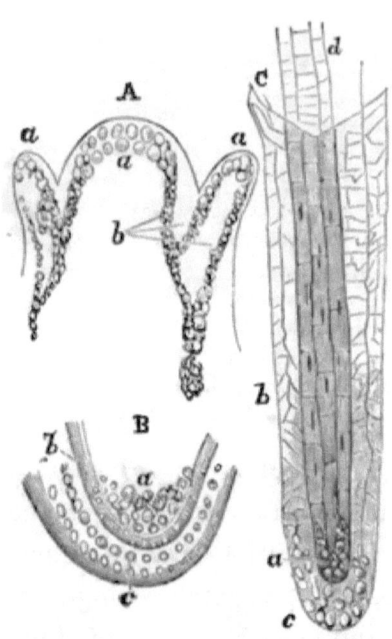

FIG. 57.—A. Mode of growth in stem. B. In root. A. *a*. Growing cells in stem, which multiply by fission. *b*. Cambium, elaborated by growing cells. B. *a*. Growing cells in root. *b*. Cells produced by growing cells. *c*. Cap, (pileorhiza.) C. Root of duckweed, (magnified.) *a*. Growing point. *b*. Root-sheath. *c*. Cap. *d*. Root.

In the growing stem the terminal cells (*a*. Fig. 57, A) multiply and enlarge, while they furnish new cells to the cambium layer. In the root, however, the multiplying cells are not quite at the extremity. A sort of cap is formed by the growing cells, (*pileorhiza*,) and receives additions to its interior, which push out the layers external to them. (Fig. 57, B, C.) Thus efficient protection is afforded to the newly-formed tissue.

The general arrangement of the tissues of a flowering plant may be seen in Fig. 58. Air passages are both intercellular and vascular, the latter in Exogens being dotted ducts and spiral vessels. The bark contains elongated *liber* or bast cells, but there are no scalariform vessels as in Acrogens. The chlorophyll (Chap. VI, Sec. 2) is found chiefly in the cells immediately under the epidermis. See also Figs. 15 and 16. The roots are supplied with water containing carbonic acid, air, and oxygen, in addition to the minerals and decompos-

ing organic matter (or *humus*) contained in the soil.
Some plants grow without attachment to the soil, deriv-
ing all their nutriment from
the air, and are called *Epi-
phytes*, (*epi*, upon; *phyton*, a
plant,) from being generally
found on trees. They differ
from true parasites, since the
latter prolong their tissues into
other plants, and prey upon
them. The Orchids may illustrate the first, and the Dodder
and Mistletoe the latter kind.

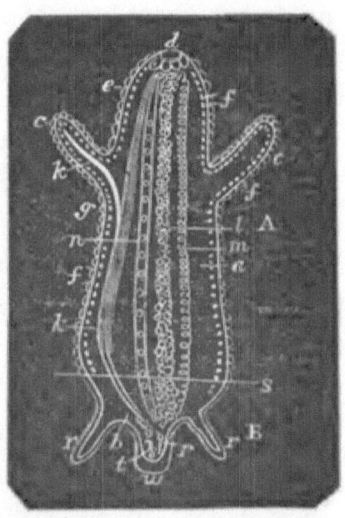

The only structure capable of
effecting the chemical changes
necessary to plant nutrition is
the chlorophyll, which is most
abundant in the leaves; hence
the materials which supply food
must be carried up to the
leaves. The ascent of fluid
from the root to the leaves

FIG. 58.—Diagrammatic Section of a
Flowering Plant, showing the different
tissues. A. Ascending axis. B. De-
scending axis. *s*. Surface of soil. *c. c.*
Appendages. *d.* Growing point of stem.
e. Epidermis. *f. f.* Stomata: *g.* Layer
containing chlorophyll, (marked by
the dotted lines.) *k. k.* Woody fiber.
l. m. n. Pith, spiral vessel, and dotted
duct—all air passages. *r. r.* Roots.
t. Growing point of roots. *w.* Cap, (pileorhiza.)

takes place by means of two distinct forces—a pushing
force, caused by absorption by the extremities of the
rootlets, and endosmose (Chap. IV, Sec. 3) from cell to
cell; and a pulling force, produced by evaporation from
the surface of the leaves.

The appendages of the root are the rootlets, and of
the stem, the leaves. Leaves are developed from the
nodes, and the internodes (Chap. VIII, Sec. 2) become
shorter toward the summit of the stem, which ends in

a terminal *bud*. Buds are also developed in the axils of the leaves, and some of them grow into branches, which repeat the characters of the stem ; but others, when the plant is fully developed, grow into stalks which support the flowers.

In Chap. IX, Sec. 11, will be found a general description of the flower, and in Chap. IV, Sec. 11, an account of the tissues forming the leaf. The arrangement of leaves and branches is also a subject of biological interest. The mode in which branches come off from the nodes gives rise to various forms of trees, such as pyramidal, spreading, or weeping. In the Italian Poplar and Cypress the branches are erect, forming acute angles with the upper part of the stem ; in the Oak and Cedar they are spreading; while in the Weeping Willow and Birch they are pendulous from their flexibility. Leaves also are placed in a fixed order for every species of plant, and this order may be expressed by an arithmetical formula. The arrangement of the leaves on the axis is called *phyllotaxis*, (*phyllon*, a leaf; *taxis*, order.) Each node of the axis may give rise to a leaf, but sometimes several nodes are approximated nearly together, and then several leaves may be produced at the same height on the stem. When two leaves are at the same level, one on each side of the stem, they are called *opposite ;* when a circle of leaves is thus produced it is called a *verticil*, or *whorl*. When a single leaf is produced at a node, and the nodes are separated, the leaves are *alternate*. The relative position of alternate leaves varies in different plants, but is tolerably uniform in each species. In a regularly-formed branch covered with leaves, if a thread is passed from one to the other, turning always in the same direction, a spiral is described, and a certain number of leaves and of complete turns occur before reaching the leaf directly above that which began the series. This may be expressed by a fraction, the numerator of which indicates the number of turns, and the denominator the number of leaves in the spiral cycle. In the Peach and Plum-tree the cycle embraces five leaves, and the spiral goes twice around the branch. This is expressed by the formula $\frac{2}{5}$. In the Alder three leaves constitute the cycle, and the spiral has only a single turn on the stem; the disposition of its leaves is represented by the fraction $\frac{1}{3}$.

In Exogenous plants the leaves are reticulated, and usually articulated to the stem. The flowers are formed upon a quinary or quaternary type; that is, their parts are in sets of fives or fours, instead of sets of threes, as Endogens. The embryo has two opposite cotyledons, or seed-lobes, which gives the term Dicotyledonous to the type.

2. According to the natural system of De Candolle, which is usually followed, Exogens are subdivided as follows :

1.) THALAMIFLORÆ, (*Thalamus*, receptacle, and *flos*, flower.) Calyx and corolla present; petals distinct, inserted into the receptacle; stamens hypogynous, or growing from below the ovary, as Ranunculus, Magnolia, Poppy, Violet, Geranium, etc.

2.) CALYCIFLORA. A calyx and corolla present, the petals distinct, but the stamens are perigynous, or attached to the calyx; as Rhamnus, the Leguminose family, the Rose family, the Syringa, the Passion-flower, Cactus, etc.

3.) COROLLIFLORÆ. Calyx and corolla present ; petals united, bearing the stamens ; as the Honeysuckle, Madder, Teazel, Composite family, Heaths, etc.

4. MONOCHLAMYDEÆ, (*Monos*, one ; *chlamus*, a cloak, or covering.) Sometimes called INCOMPLETÆ. Corolla wanting ; a calyx or simple perianth present. (Even this sometimes absent.) It is divided into two sections :

A. *Angiospermæ*. Seeds contained in an ovary, as Amaranth, Phytolacca, Buckwheat, Laurel, Begonia, Nettle, Fig, and the Catkin-bearing family.

B. *Gymnospermæ*. Seeds naked. Their woody tissue is marked by disks, (Chap. IV. Sec. 9 ;) as the Coniferæ and Cycas family.

3. Among the INCOMPLETE Exogens, belonging to the section of *Gymnosperms*, or naked-seeded Exogens, are found the Cycas family, (CYCADACEÆ,) which greatly resemble the Palms and Tree-ferns, and the Cone-bearing family, (CONIFERÆ,) divided into the Fir and Spruce tribe, (*Abietineæ*,) the Cypress tribe, (*Cupressineæ*,) the Yew tribe, (*Taxineæ*, and the Joint-fir tribe, (*Gnetaceæ*.)

The Coniferous plants are noble trees or evergreen shrubs, and furnish valuable timber and other important products, as turpentine, pitch, and resin. The Pine (Fig. 59) is one of the most perfect trees of the forest, considered in respect to its beauty and uses. " Its character and glory," writes Mr. Ruskin, "consist in its right doing of its hard duty, and forward climbing into those spots of forlorn hope where it alone can bear witness of the kindness of the Spirit that cutteth out rivers among the rocks, as it covers the valleys with corn ; and there, in its vanward place, and only there, where nothing is withdrawn for it, nor hurt by it, and where nothing can take part of its honor, nor usurp its throne, are its strength, and fairness, and price, and goodness in the sight of God to be truly estimated."

4. Among the ANGIO-SPERM EXOGENS, or those of the Incomplete class whose seeds are inclosed in an ovary. The principal families are the Marvel of Peru, the Amaranth, the Phytolacca, the Buckwheat, the Begonia, the Laurel, the Nutmeg, the Oleaster, the Daphne, the Sandalwood, the Birthwort, the Pitcher-plant, the Rhizogen, the Spurge, the Nettle, the Pepper, the Walnut, and the Catkin-bearing families.

To the Buckwheat family (POLYGONACEÆ) belong the

FIG. 59.--Pines, (*Pinus Sylvestris.*)

Buckwheat, (*Fagopyrum esculentum,*) the Sorrel, (*Rumex Acetosa,*) and Rhubarb, (*Rheum palmatum.*) In the Laurel family are the Laurel, Bay, Camphor, Sassafras, and Cinnamon trees.

The Pitcher-plants (*Nepenthes*) are among the curiosities of the vegetable world, on account of the pitcher

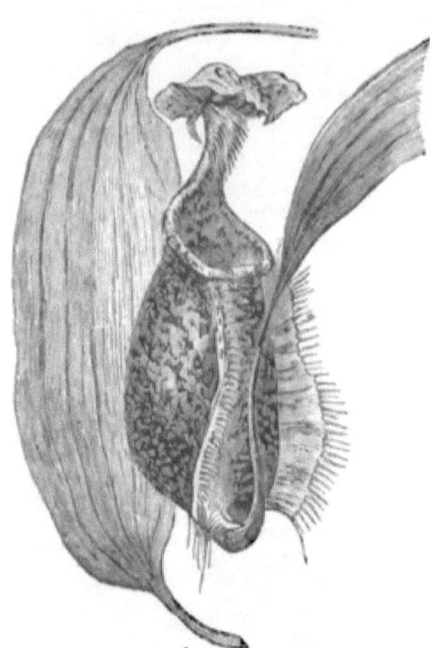

formed at the end of the leaf. This is furnished with a lid, and contains a limpid fluid secreted by glands in the cavity, and in sufficient quantity to drown flies and other insects which fall into it. (Fig. 60.) Since the publication of Mr. Darwin's "Insectivorous Plants" these secreting leaves (together with those of several other species) have attracted much attention, as in all probability there is in

FIG. 60.—Pitcher-plant, (*Nepenthes distillatoria.*)

these arrangements provision for a true digestion, as in the case of animals.

The *Euphorbiaceæ*, or Spurge family, contains many trees, shrubs, and herbs, abounding in acrid milky juice, which is generally poisonous. *Siphonia elastica* is one of the plants which supplies caoutchouc, or India-rubber. The seeds of *Croton Tiglium* affords Croton oil, and those of *Ricinus communis* (or *Palma Christi*) furnish Castor-oil. In the root of *Janipha Manihot* there is much starchy matter mingled with a volatile poison. The latter is removed by heat or washing, and the starch is used as Cassava bread. Tapioca and Brazilian Arrow-root are said to be procured in this way.

In the NETTLE family are about six hundred species, including the common Nettle, Hemp, Hop, Elm, Fig, Mulberry, Bread-fruit, the Banyan, (*Ficus indica*,) etc.

The Catkin-bearing family (AMENTACEÆ) is the largest and most important of this order, since it contains all the most important timber trees. (Fig. 61.) The

Fig. 61.—Woodland Scenery.

Alder, Birch, Willow, Poplar, Oak, Chestnut, Hornbeam, and Plane are here brought together because of the similarity of their fructification. They produce flowers of one sex only, the males of which are in catkins, in which the

13

flowers have neither calyx nor corolla, but merely a single scale. Their bark has an astringent quality from the
presence of tannin, and some, as the Willow, yield a
valuable tonic febrifuge, (*Salicin.*) The fruit of many
species contains starchy matter, rendering it edible by
man or animals, as the acorns of oak, mast of birch, nuts
of the hazel, etc.

5. In the order COROLLIFLORÆ, or Exogens having
the petals united, and bearing the stamens, are to be
found the Mistletoe, the Honeysuckle, Peruvian bark,
Valerian, Teazel, Harebell, Lobelia, Heath, Cranberry,
Ebony, Holly, Jasmine, Olive, Asclepias, Dog-bane,
Gentian, Trumpet-flower, Phlox, Convolvulus, Borage,
Nightshade, Figwort, Labiate, Vervain, Acanthus, Prim-
rose, and Composite families.

The Honeysuckle family (CAPRIFOLIACEÆ) is divided
into the true Honeysuckle tribe (*Loniceræ*) and the El-
der tribe, (*Sambuceæ.*)

The Peruvian Bark family (RUBIACEÆ) contains, in
addition to the Peruvian bark of commerce, (*Cinchona,*)
the Ipecacuanha, (*Cephaelis Ipecacuanha,*) the Coffee-tree,
(*Coffea arabica,*) and the Madder, (*Rubia tinctoria.*)

The Heaths (ERICACEÆ) contain many beautiful and
showy plants, as the Rhododendrons, Azaleas, and Kal-
mias. The Partridge-berry, (*Gaultheria procumbens,*) the
Bear-berry, (*Arctostaphylos Uva-Ursi,*) and the Chima-
philla, (*Pyrola umbellata,*) are sometimes used in medicine.

In the Olive family (OLEACEÆ) is placed the Olive,
Lilac or Syringa, and the Ash, (*Fraxinus excelsior.*)

The Gentians (GENTIANACEÆ) are mostly dwarf her
baceous plants, with deep blue flowers.

The CONVOLVULACEÆ, or Bind-weed family, are twining plants with showy flowers, except the tribe of Dodders, (*Cuscuteæ*,) which are leafless parasites. Here we find the Jalap, and Scammony, and Sweet-Potato, (*Batatas edulis*.)

The Nightshade family (SOLANACEÆ) contains the Potato, (*Solanum tuberosum*,) the Deadly Nightshade, (*Atropa Belladonna*,) the Henbane, (*Hyoscyamus niger*) the Thorn-apple, (*Datura Stramonium*,) Tobacco, (*Nic.-tiana Tabacum*,) Cayenne-pepper, (*Capsicum annium*,) the Tomato, (*Lycopersicum esculentum*,) etc.

The LABIATÆ are characterized by two long and two short stamens, four little nuts or naked seeds, and irregular corollas. The plants are generally fragrant and aromatic, and none of them are injurious. Many are used in medicine as carminatives. Mint, Lavender, Sage, Savory, and Balm, are examples of the family. From Thyme a sort of camphor has been procured called Thymol, which has similar antiseptic properties to Carbolic acid, but with pleasant odor.

The family COMPOSITÆ is a very extensive one. The florets are arranged in involucrated heads, and the anthers cohere into a cylinder. It is subdivided into three sections: 1. *Cynarocephalæ*, (from *cynara*, the Artichoke,) having all the flowers tubular; the involucre, hard, conical, and often spiny, as the Thistle, Burdock, etc. 2. *Corymbiferæ*, (*corymbus*, a comb, and *fero*, to bear,) having tubular florets in the disk (center) and ligulate in circumference, (or ray;) involucre hemispherical, leafy, or scaly, rarely spiny, as Feverfew, Wormwood, Tansy, Arnica, and Sunflower. 3. *Cichoraceæ*, (*cichorium*, succory,) having the florets all ligulate, as Chicory, Dandelion, and Lettuce. The Daisies, Asters, Chrysanthemums, and Dahlias of the gardens are all composite flowers.

6. In the subdivision of CALYCIFLOREÆ are placed Exogens which have a calyx, and corolla with distinct petals, and whose stamens are attached to the calyx.

In the Buckthorn family (RHAMNACEÆ) we find the genus Rhamnus, several of whose species yield cathartic medicine, and Ceanothus, or Mountain tea.

The Cashew-nut family (ANACARDIACEÆ) contains the Cashew-nut, (*Pistacia vera*,) *Rhus Toxicodendron*, or Poison-oak, and many plants which furnish varnishes, as the Japan lacquer, (*Stagmaria verniciflua*.)

A number of fragrant balsamic resins, including myrrh, (*Balsamodendron Myrrha*,) are obtained from plants of the Amyris family, (AMYRIDACEÆ.)

The Pea and Bean family (LEGUMINOSÆ) is very extensive, containing more than four hundred and fifty genera and six thousand five hundred species. It embraces many valuable medicinal plants, as those yielding Senna, Gum-arabic, Tragacanth, Catechu, and Kino; important dyes, as Indigo and Logwood; valuable timber-trees, as Locust-tree and Rosewood; and plants furnishing nutritious food, as the Bean and Pea. This order has been divided into three sub-orders, 1. *Papilionaceæ;* having papilionaceous flowers, the petals imbricated in æstivation, and the upper one exterior. The plants of this section often have beautiful showy flowers, as Robinia, Laburnum, Lupinus, etc. The various kinds of Clover, Beans, Peas, and Pulse belong to it. The *Glycyrrhiza glabra*, or plant yielding liquorice-root, the

Myroxylon peruiferum, or source of the Balsam of Peru, and many other plants having medicinal qualities, are found here. 2. *Cæsalpineæ.* Flowers irregular, but not papilionaceous, petals spreading, imbricated in æstivation, upper one interior. Here we find the place of several plants used in medicine, as various species of *Cassia* or Senna, the Tamarind-tree, and the Logwood, (*Hæmatoxylon.*) 3. *Mimoseæ.* Flowers regular, petals valvate (without overlapping) in æstivation; as the different species of *Acacia*, yielding Gum Arabic, and the *Mimosæ,* or Sensitive plants.

The Rose family (ROSACEÆ) is also a very large one belonging to the *Calyciftoreæ.* Its sub-orders are, 1.) *Chrysobalanceæ*, petals and stamens irregular, ovary stipitate, its stalk adhering to the side of the calyx, style basilar, fruit a 1–2-celled drupe, (or fleshy fruit.) 2.) *Amygdaleæ*, tube of calyx lined with a disk, styles terminal, fruit a drupe. 3.) *Spiræeæ*, calyx-tube herbaceous, lined with a disk, fruit of numerous follicles, seeds apterous. 4.) *Quillaieæ*, flowers unisexual, calyx-tube herbaceous, fruit capsular, seeds winged at the apex. 5.) *Sanguisorbeæ*, petals none, tube of calyx thickened and indurated, stamens definite, nut solitary, inclosed in the calycine tube. 6.) *Potentilleæ*, calyx-tube herbaceous, lined with a disk which sometimes becomes fleshy, fruit consisting of numerous achænia, (small, brittle, seed-like fruit.) 7.) *Roseæ*, calyx-tube contracted at the mouth, becoming fleshy, lined with a disk, and covering numerous hairy achænia. 8.) *Pomeæ*, tube of calyx more or less globose, ovary fleshy and juicy, lined with a thin disk, fruit a 1–5-celled, or spuriously 10-celled, pomum.

13*

Many of the plants of this order yield edible fruits, as Raspberries, Strawberries, Plums, Apples, Pears, Cherries, Peaches, and Apricots. Plants of the sub-order Amygdaleæ are remarkable for the presence of hydrocyanic acid, as in the kernel of the Almond, (*Amygdalus communis*,) especially the bitter Almond; the leaves of the Peach, (*Amygdalus persica*,) and of the Cherry-laurel, (*Prunus Laurocerasus*.) The sub-order Pomeæ supplies Apples, Pears, and Quinces. The seeds contain hydrocyanic acid. The other sub-orders have plants distinguished by astringent and tonic properties, as the root of *Potentilla Tormentilla*, and the petals of *Rosa gallica*, the Red Rose.

Fig. 62.—A Mangrove Forest.

The RHIZOPHORACEÆ, or Mangrove family, is named after *Rhizophora Mangle*, the Mangrove, which forms thickets at the muddy mouths of rivers in tropical countries, and sends out adventitious roots which often raise up the main trunk, and give the tree the appearance of being supported on stalks. (Fig. 62.) The fruit is sweet and edible.

The Myrtle family (MYRTACEÆ) contains trees or shrubs which are usually natives of warm countries. Some of the genera are peculiar to Australia, as the Eucalyptus, or Blue Gum-tree, which is being planted extensively in California. It is a rapid grower, and promises to be serviceable as a forest tree. It contains a medicinal balsamic resin. The Pimento, (*Myrtus Pimenta*,) the Pomegranate, (*Punica Granatum*,) and various species of edible Guavas and Rose-apples belong to this order.

The Evening Primrose (*Œnothera*) and the *Fuchsia* belong to the order ONAGRACEÆ, or the Evening Primrose family.

The Cucumber family (CUCURBITACEÆ) contains many plants that are drastic purgatives, and others whose fruits under cultivation are edible, as the Melon and the Colocynth, both species of the same genus, (*Cucumis*.)

The Passion-flowers (PASSIFLORACEÆ) received their name from a fancied resemblance to the scenes at Calvary. The superstitious monks saw in the five anthers a resemblance to the wounds of Christ; in the triple style, the three nails on the cross; in the central pillar, the cross itself; and in the filamentous processes, the rays of light round the Saviour, or the crown of thorns.

The PORTULACACEÆ, or Purslane family, are chiefly herbaceous plants, found in dry, barren situations, or on the sea-shore. Some of them have tuberous roots which have been proposed as substitutes for the potato, as *Claytonia tuberosa*, and *Melloca tuberosa*. The first is a Siberian plant, the other a native of Peru.

The Cactus family (CACTACEÆ) contains many succulent plants, destitute, for the most part, of leaves, the place of which is supplied by fleshy stems of grotesque figures. Some are angular, and grow to a height of

FIG. 63.—A Group of Cactaceæ.

thirty feet; others are roundish, covered with stiff spines, and not over a few inches high. Their flowers are often large and beautiful, varying from pure white to rich scarlet, or purple. Some are night-flowering, as the *Cereus grandiflorus*. In Mexico and Southern California there are a large number of species, some of which are of gigantic size. (Fig. 63.)

The Gooseberry and Currant family, (GROSSULARIA-CEÆ,) the Saxifrage family, (SAXIFRAGACEÆ,) the Witch-hazel family, (HAMAMELIDACEÆ,) are all of this section of Exogens, with many others. The Umbelliferous family (UMBELLIFEREÆ) are characterized by the radiating or umbrella-like arrangement of the florets. The properties of the plants of this family are various. Some yield articles of diet, as the Parsnip, (*Pastinaca sativa*,) Carrot, (*Daucus Carota*,) and Parsley, (*Petroselinum sativum.*) Others yield milky juices, which dry into a fetid gum-resin, as the *Ferula Assafœtida*, yielding Assafœtida, and *Dorema Ammoniacum*, which produces Ammoniac. Others again supply a carminative and aromatic oil, as Caraway-seeds (*Carum Carui*) and Fennel, (*Fœniculum dulce.*) Some species are quite poisonous, as the *Conium maculatum*, or Hemlock, which contains a volatile alkaline poison, called Conia.

7. In the sub-class, or order THALAMIFLORÆ, the stamens are inserted under the pistil into the thalamus, or receptacle. The petals, also, are inserted into the receptacle. In some cases the petals are abortive, and it becomes hard to determine whether the plant belongs to this division or to Monochlamydeæ.

The RANUNCULACEÆ, or Crowfoot family, is characterized chiefly by having several distinct carpels, above numerous stamens. The plants are generally narcotic acid poisons. The Ranunculus, Anemone, Larkspur, Aconite, and Peony are examples.

The leaves of *Aconitum Napellus*, or Monkshood, are used in medicine, as well as the rhizome of *Podophyllum peltatum*, or May Apple.

The Poppy family (PAPAVERACEÆ) differs from the last in having the carpels united into an undivided ovary,

and in having milky or colored juice. Opium is the dried juice of *Papaver somniferum*, (Fig. 64,) or Poppy, and its varieties. The Celandine (*Chelidonium majus*) yields an orange-colored juice, which is said to be acrid. In the leaf of this plant may be seen under the microscope the movement of the sap in the laticiferous vessels. *Sanguinaria canadensis*, or Blood-root, has emetic and cathartic properties. The yellow California Poppy (*Eschscholtzia*) is remarkable for the two sepals of its calyx adhering at the edge, and separating at the base by the growth of the flower, so as to form a sort of calyptra, or hood,

FIG. 64.—The Opium-Plant, (*Papaver somniferum*.

over the unexpanded petals, resembling the extinguisher of a candle.

MAGNOLIACEÆ, the Magnolia family, contains the well-known Magnolias, remarkable for large odoriferous flowers, the Swamp Sassafras, (*M. glauca*,) whose bark is used as a substitute for the Peruvian bark, and the *Liriodendron tulipifera*, or Tulip-tree, etc.

The Side-saddle family (SARRACENIACEÆ) contains the genera *Sarracenia* and *Darlingtonia*, which (like Nepenthes) are characterized by a pitcher-like appendage to the leaf, containing a fluid secretion, supposed to have the power of digesting insects which fall into it.

CRUCIFERÆ, the Cruciferous, or Cress-wort family, known so readily by their four cruciate petals, contains a large number of plants, none of which are poisonous, although some are stimulant and even acrid. Most of the common culinary vegetables belong to this order, as Cabbage, Cauliflower, Turnip, Radish, Cress, and Mustard. Many garden flowers also are of this family, as Wallflower and Alyssum.

The Violet family, (VIOLACEÆ,) the Mignonette family, (RESEDACEÆ,) the Berberry family, (BERBERIDACEÆ,) the Rock-Rose family, (CISTACEÆ,) the St. John's-wort family, (HYPERICACEÆ,) the Vine family, (VITACEÆ,) the Geranium, or Crane's-bill family, (GERANIACEÆ,) the Wood-sorrel family, (OXALIDACEÆ,) and many others, must be passed by, since space forbids us to enlarge.

The Quassia family (SIMARUBACEÆ) is noted for the bitter and tonic principle contained in the wood of *Quassia amara*, and other species.

The Rue family (RUTACEÆ) is also known in medicine, since it furnishes Rue, Buchu, (*Barosma crenata*,) and other agents.

The Flax family (LINACEÆ) furnishes the well-known Flax, (*Linum usitatissimum*,) whose inner bark yields linen and cambric. The seeds are mucilaginous and oleaginous.

The Water-lily family (NYMPHÆCEÆ) contains plants with showy flowers. (Fig. 65.) *Victoria regina* is one of

FIG. 65.—Common Water Lily, (*Nymphæa alba.*)

the largest known, the white and rosy flowers being four feet in diameter, and the leaves fifteen feet across, according to Schlieden.

DROSERACEÆ, the Sundew family, is remarkable for its insectivorous properties. The *Droseras* are furnished with glandular hairs, which exhibit drops of fluid in sunshine, hence the name.

Dionæa muscipula, Venus's Fly-trap, has the laminæ of the leaves in two halves, each furnished with three irritable hairs, which, on being touched, cause the folding of the divisions in an upward direction.

The Chickweed and Pink family (CARYOPHYLLACEÆ) contains all the Carnations, or Pinks, (*Dianthus*,) Chickweed, (*Stellaria media*,) etc.

The Mallow family (MALVACEÆ) contains many wholesome mucilaginous plants. The Mallow, (*Malva*,) the Hollyhock, (*Althæa rosea*,) the Abutilon, (*A. esculentum*,) and the Cotton-plant, (*Gossypium*,) belong here. (Fig. 66.)

The produce of the latter plant employs the labor of a million and a half of people in England alone, and furnishes clothing to hundreds of millions.

The Tea family (TERN-STRÆMIACEÆ) has in it the beautiful Camellias of Japan, and the plants which furnish tea, (*Thea-viridis* and *Bohea*.) The use of the leaves of these plants is immense, no less than fifty-six millions of pounds being imported into Great Britain in a single year, (1846.)

FIG. 66.—The Cotton-Plant, (*Gossypium.*)

The bitter principle in tea, called *theine*, may be procured by adding a slight excess of acetate of lead to a decoction of tea, filtering hot, evaporating, and subliming.

The Orange family (AURANTIACEÆ) contains about a hundred species. The plants contain receptacles of volatile oil. The fruit has an acid or subacid pulp, and the wood is compact. The Orange, Lemon, Lime, Citron, and Shaddock belong here.

ACERACEÆ, the Maple family, contains the Maple and Sycamore, (*Acer pseudo-platanus.*) The Sugar Maple (*A. saccharinum*) yields a sap from which sugar is manufactured.

The Mahogany family (CEDREIACEÆ) contains plants with an aromatic fragrance. *Swietenia Mahogoni* sup-

14

plies the well-known mahogany wood, and *Chloroxylon Swietenia*, satin wood.

8. In the rapid sketch we have made of the vegetable kingdom, we have omitted the minute botanical details characteristic of each family, and have only given the principal differences and resemblances of types and classes, with some few representative forms in the most important families. These general peculiarities of plants serve in a great degree to define the character of landscape scenery in various parts of the world. Gray and withered Lichens clothe the barren confines of vegetation toward the snow-line of mountains or at the north, while Mosses form a silken cushion over rock and soil with their delicate leaflets.

Grasses are characterized by their sociability, and call forth agreeable sensations by their soft carpet of green and pliant leaves. The Sedges, on the contrary, with stiff and rugged stems and leaves, rejected by cattle, awaken no pleasing associations. In tropical climates, as in Hindustan, the tall Bamboo sometimes overtops the trees, and forms a meadow above the forest. There the Plantain stem swells with sap, the leaves expand and are split by the wind, and the great flower-bunches beam with intense color. Between the reeds and the banana plants the lilies may be placed. The arrow-shaped leaves of the Aroids, with strange and often brightly-colored spathes, mark the transition to the Orchids.

The stems as well as the leaves of plants often give character to a landscape, as in the Heaths—low, branching, dull-green or gray shrubs, whose blossoms scarcely obliterate the melancholy impression produced where

they abound. The arborescent Heaths (*Casuarinæ*) form
many of the gloomy woods of Australia. Still more
striking are the forms of the thorny Cactuses, (Fig. 63,)
consisting merely of fleshy stems and branches of singular shapes. The Yuccas of Mexico, the great African
Aloes, and the Grass-trees of Australia, with their solid
liliaceous leaves, of a dull green, afford a picture of immovable repose. The stiff, shining leaves of Pandanus,
or Screw Pine, arranged in spiral lines, contrast greatly
in the Sandwich Islands with the finely divided leaves
of the Fern, spreading in graceful elegance, and trem-
bling in the breeze. Between these two extremes is the
Palm-form, which gives most characteristic beauty to the
tropical world. Some Palms have feathered leaves,
others have fan-leaves, and in some of the umbrella
Palms the crown consists of a few fans elevated on long,
slender stalks. In all the inflorescence breaks from the
stem below the origin of the leaves, and the sheath
hangs down, often several feet long. The shape and
color of the fruit varies from the large triangular Cocoa-
nut to the berry of the Date. The aerial summits of
the Palms, projecting like a colonnade above the thicket,
and crowned with leaves, give them an air of beautiful
majesty. (Fig. 51.) Deciduous, or Leafy woods, (Fig.
61,) with their branching stems and broad foliage, form
dense, compact, vegetable masses, characteristic of temperate climes. Wand-like forms, with narrow, fluttering
leaves, often covered with silvery down on the under
side, are represented by the Willow and Poplar, and in
the south of Europe by the Olive. The Conifers, or
Needle-leaved woods, are distinguished by their narrow,

dark-green leaves and whorl-like branches. (Fig. 59.)
In the tropical or equinoctial regions the mass of leafy
woods is marked by the prevalence of the Mallow-form,
with long-stalked and palmately-lobed leaves. The
giant Baobab, the barrel-like trunk of the Bombax, and
the purple-blossomed Hibiscus bush belong to this class.
The Australian Laurels and Myrtles are allied to the
northern Willows, yet their rigid leaves, shining as if
varnished, or covered with a silvery felt which mingles
with the shining green, give them a characteristic phys-
iognomy.

Thus even a general observer may notice variety
enough to indicate that a free intelligence has arranged
these forms to minister mental enjoyment, as well as to
supply the needs of intelligent creatures. Archbishop
Trench has well said that the characters of nature which
every-where meet the eye "are not a common, but a sacred writing—they are hieroglyphics of God."

CHAPTER XI.

PROTOZOA.

Since all bioplasm possesses certain common characters, and the bioplasm of one plant or animal produces formed matter of a very different
kind from that resulting from another portion of bioplasm, we must admit that in nature there are different kinds of bioplasm.—DR. BEALE'S
Bioplasm.

1. IN studying the structure of the Protozoa, or primitive animals, we seem to be going backward, since each
is composed of a single mass of bioplasm, like the simplest vegetables, or Protophytes. Although similar in
structure, the Protozoa and the Protophytes are biolog-
ically distinct in function, since the latter generally de-
compose Carbonic acid under the influence of light, and
generate Chlorophyll and albuminous compounds in a
manner similar to the leaf-cells of the most perfect
plant, while the Protozoa ingest and digest both animal
and vegetable food as effectively as the most complex
animals.

We have already seen (Chap. II, Sec. 7) that all living
matter, or bioplasm, has essentially spontaneous motion,
nutrition, growth, and reproduction. We cannot con-
ceive, therefore, of any form of life, either vegetable or
animal, without these characteristics. The simplest and
most embryonic structures in both kingdoms of nature
exhibit these functions. Whether spontaneous motion
is proof of consciousness and will, will be considered
hereafter.

14*

2. The MONERA of Prof. Haeckel, if the group shall be accepted by naturalists, will include the simplest proto-zoans, or those in which the entire living body is a mere particle of bioplasm, without nucleus, vacuole, invest-ment, or other structure, yet capable of bioplasmic mo-tions and other functions. *Bathybius*, referred to in Chap. IV, Sec. 3, was supposed to be of this class.

3. The GREGARINIDÆ are parasitic. Each consists of a single cell, which passes through changes similar in many respects to Protophytes. It becomes globular and encysted in a horny envelope, and the inclosed bioplasm breaks up into particles which become " pseudo-navicel-læ," or forms similar to the *Naviculæ* of the family Dia-tomaceæ. It is not unlikely that the *Gregarinæ* are but phases in the life-history of other parasitic worms.

4. RHIZOPODA. The Rhizopods, or root-footed Pro-tozoans, (*rhiza*, a root, and *pous*, foot,) are characterized by the power of spontaneously throwing out delicate processes of their bioplasm, called *pseudopodia*, or false feet, for prehension or locomotion. They have no cilia. Dr. Carpenter has divided the class into three orders: 1. *Reticularia*, whose bodies are indefinite extensions of viscid bioplasm, freely branching and subdividing into fine threads, but readily coalescing when they come into contact. 2. *Radiolaria*, whose bioplasm has an invest-ing membrane of formed material which prevents the coalescence of the radiating or rod-like extensions of the pseudopodia. 3. *Lobosa*, whose bioplasm has an in-vesting membrane, or ectosarc, and whose false feet are lobose extensions of the body itself.

In the first order, that of reticularian rhizopods, we

find many genera and species which secrete a shell or external envelope of Carbonate of Lime, or Chalk. These shells are often of singular beauty. They are generally perforated by a large number of minute openings for the passage of the pseudopodia, and hence are termed *Foraminifera*, (*foramen*, an aperture ; *fero*, I carry.) (Fig. 67.) Some of these foraminifera are single chambers, often like striated flasks, (*Lagena*,) but

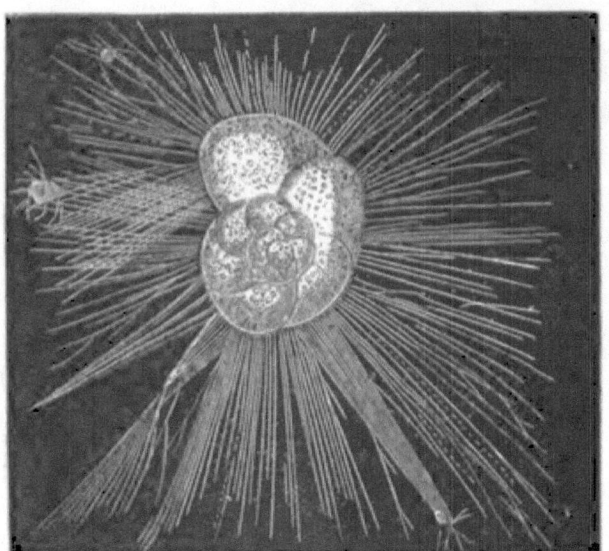

FIG. 67.—Rosalina ornata, with its pseudopodia extended.

others are compound, either straight, (*Nodosaria*,) spiral, (*Rotalia*,) or irregular, (*Globigerina*.) These shells are generally microscopic, although some, as the *Nummulites*, may be an inch in diameter, and the fossil *Eozoon Canadense*, which is referred to this order, was of indefinite size. The Foraminifera accumulate in the bed of the ocean in great numbers, yet in former ages they were still more prolific, since the Chalk cliffs of England.

the building-stone of Paris, and the limestone of the Egyptian pyramids, are composed of their remains.

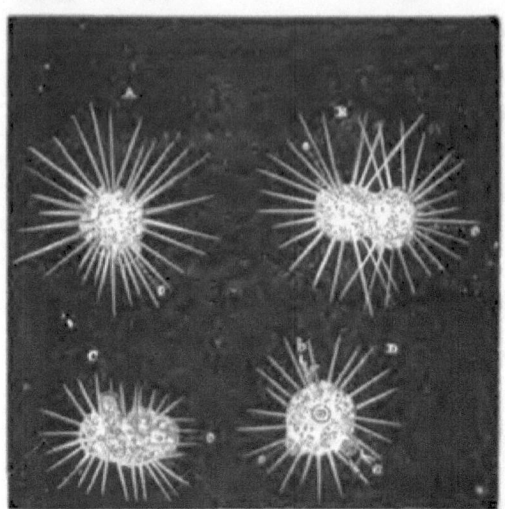

FIG. 68 —Actinophrys sol, in different states. A. In its ordinary sunlike form, with a prominent contractile vesicle, *o.* B. In the act of division or of conjugation, with two contractile vesicles, *o, o.* C. In the act of feeding. D. In the act of discharging fæcal (?) matters, *a* and *b.*

FIG. 69.—A. Podocyrtis Schomburgkii. B. Rhopalocanium ornatum.

In the Radiolarian order is placed the *Actinophryssol,* (Sun-Animalcule,) (Fig. 68;) many species of *Polycystina.* which secrete silicious shells, of various shapes and of wonderful beauty, (Fig. 69;) and colonies of gelatinous rhizopods, (*Thalassicollida,*) containing silicious spicules.

To the order Lobosa belongs the *Amœba princeps,* (Fig. 2,) to which reference has been so often made, since it has occupied so important a position in modern biology. Chap. II, Secs. 2, 3, 5.

5. INFUSORIA, or Animalcules. The term Infusoria is applied to

this class because the species abound in any infusion of vegetable or even animal matter which is allowed to putrefy. The word was formerly applied to a much larger number of species than now, since many forms once considered animal have been placed in the vegetable kingdom, as the Desmids, the Diatoms, the Volvox, and many other Protophytes. The Rotifers, or wheel-animalcules, also, on account of their organization, are referred to the articulate type of animal life. It is possible that some of the Infusoria may be but larval forms of higher animals. After all this pruning, however, the class is still a large one, and full of interest. It is divided into three orders: *Ciliata*, *Suctoria*, and *Flagellata*.

Ciliated Infusoria (*ciliata*) are most numerous, and are named from the cilia, or hair-like organs, round the mouth, or body, of the animalcule. Cilia are not confined to animalcules. They are found among Protophytes, (Chap. VI, Sec. 3.) They also exist in many organs of the higher animals, as in the respiratory passages even of man himself. They appear to be tapering prolongations of bioplasm, or of formed material in connection with bioplasm, and have a sort of waving or circular motion. In the internal organs of man their actions are constant, entirely without consciousness, and may continue long after the death of the body. In the animalcules the ciliary action is interrupted and renewed in such a way as to impress an observer with the idea of choice and direction.

Vorticella, or the bell-shaped animalcule, was described in Chap. I, Sec. 6, and the life-history there given may serve for a representation of the entire order.

Epistylis differs from Vorticella in having a branching and non-contractile stem.

Vaginicola possesses a horny, cuticular case, (a *carapace*, or *lorica*,) into which the animal can retire.

Stentor is a fresh-water infusorian, shaped like a trumpet.   It may be found either free or attached.

Paramecium, (Fig. 70,) is a free, fresh-water animal-

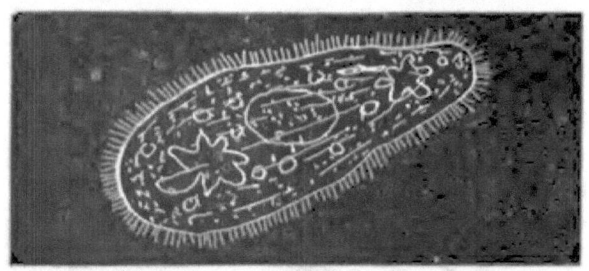

FIG. 70.—Paramecium Aurelia, an Infusorian animalcule, magnified 300 times.

cule, shaped like a slipper, the hole for the foot being represented by the mouth.

Suctorial Infusoria (order *Suctoria*) may be illustrated by the parasitic *Acineta*, Chap. I, Sec. 6.   They have filaments ending in suctorial disks, which are capable of protrusion and retraction, and are used for prehension.

The Flagellate Infusoria (order *Flagellata*) perform locomotion by means of long filaments, or flagellæ, which may be single, double, or multiple.

The *Noctiluca*, (Fig. 71,) is the best-known member of this order.   It is very minute, about one eightieth of an inch in diameter, and presents little more structure under the microscope than a simple sac of bioplasm, with vacuoles, an oral aperture, and a tail of flagellum, but at night these tiny beings light up the ocean with myriads of lamps, whose phosphorescent property is yet a profound mystery.

The *Cercomonad*, an animalcule, with a long flagellum at each end, is noted for the thorough investigations made by Messrs. Döllinger and Drysdale. These gentlemen found it would increase for several days by fission. Then it would lose the

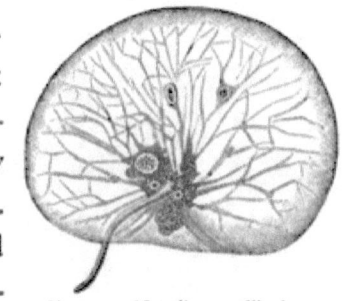

Fig. 71.—Noctiluca miliaris.

flagellæ and assume an amœboid form. Two of these amœbiform Cercomonads would conjugate and become encysted, and the rupture of the cyst gives exit to minute germs, which grow into the original parent form. A temperature of 150° F. sufficed to destroy the adult forms, but at 300° F. the germs still lived and developed. This latter fact makes strongly against the theory of spontaneous generation.

6. SPONGES, (*Spongida.*) What we familiarly call a sponge is but the skeleton of a colony of Protozoa. In this class a number of bioplasts, whose individuality is still almost if not complete, are united together, supported on a skeleton of horny, silicious, or calcareous fibers united so as to form a net-work of tubes.

In a living sponge currents of fluid set in through minute pores on the surface, and come out in large streams through the larger apertures, (*oscula.*) These currents are kept up by the cilia connected with the bioplasmic masses which line the canals and cover the skeleton. By means of these currents particles of food

are brought within reach of the bioplasts. (Fig. 72.)
The Sponges are divided into three orders: Horny,

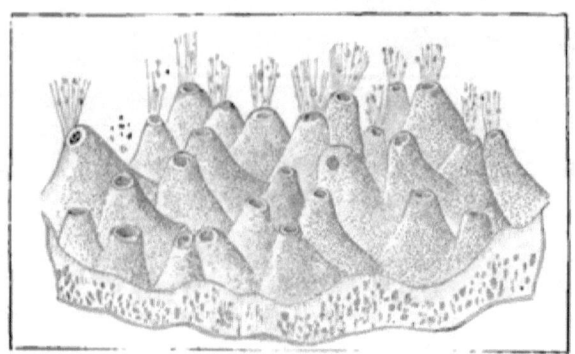

FIG. 72.—Sponge in action.

Silicious, and Calcareous sponges. In the first order
(*Keratosa*) is found the sponge of commerce, which owes
its value to the fineness of its fibers and the absence of
silicious spicules. Some sponges of this order have

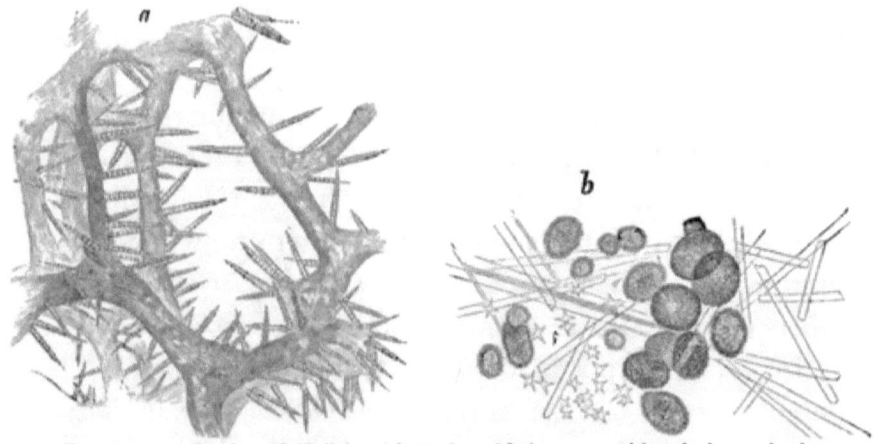

FIG. 73. — *a*. Portion of Halichondria (?) from Madagascar, with spicules projecting
from the fibrous network. *b*. Silicious Spicules of *Pachymatisma*.

flinty spiculæ of various shapes, as pins, clubs, crosses,
hooks, and anchors. (Fig. 73.) The silicious sponges

(*Silicea*) sometimes have their spicules woven or fused together, as in the beautiful *Euplectella*, or Venus's Flower-basket. In *Hyalonema*, the glass-rope, the long spicules, are twisted together.

In the order *Calcarea* the skeleton is composed of Carbonate of lime. Except a few fresh-water species, as *Spongilla*, sponges are marine. The best sponges of commerce are from the Mediterranean.

7. The colonies of bioplasts in Thalassicollida and in Sponges are analogous to the higher types of animal life, yet the individual cells are so loosely bound together, and so capable of living and performing all their functions apart, that they are ranked as Protozoa, as the colonies of Volvocineæ, Nostochaceæ, and Confervaceæ, are placed among the Protophytes.

8. The essential difference in the vital powers of different classes of living things, and of the individuals of each class, is well exhibited in the following passage from Johnston's " British Sponges :" " For example, it is very common to find growing on the same rock, or sea-weed, a silicious, a calcareous, and a horny sponge ; they have all the same exposure, and are all recipients of the same nutriment, yet does each act upon this differently. One extracts from the fluid silica, which it causes to assume a solid crystalline form ; another selects in the same manner the calcareous particles, which, obedient to the laws of life, assume figures novel to them in their mineral state ; and again, another rejects both the lime and the flint as injurious to its constitution."

15

CHAPTER XII.

RADIATA.

If we are astonished that so great deeds should proceed from the little and low, it is because we fail to appreciate that little things, even the least of living or physical existences in nature, are, under God, expressions throughout of comprehensive laws, laws that govern alike the small and the great.—Dana, *Corals and Coral Makers.*

1. In the simple Protophytes and Protozoa we find the essential structure to be a single cell, or mass, of bioplasm, having in one vegetable and in the other animal characteristics. In some instances there is a colony, or association, of bioplasts, with certain mutual relations; but as each bioplast is but loosely connected with the others, and is capable of living and performing all its functions while in a state of independence, these colonies are conveniently considered among primordial types. In the higher forms of life, either animal or vegetable, each individual is composed of many bioplasts, derived by subdivision of the primitive mass. With the division of the structure there is also a differentiation of function, so that no bioplasts of the structure, save those which are appropriated to reproduction, can normally pursue an independent existence.

2. The Radiate type of animal life is characterized by the idea expressed in the word *radiation.* "In Radiates we have no prominent bilateral symmetry, such as exists in all other animals, but an all-sided symmetry, in

which there is no right and left, no anterior and posterior extremity, no above and below. It is true that in
some of them there are indications of that bilateral symmetry which becomes a law in the higher animals ; but
whenever such a tendency is perceptible in the Radiates
it is subordinate to the typical plan on which the whole
group is founded." *

3. Radiate animals are subdivided into I. CŒLEN-
TERATA, or Cœlenterates, (koilos, hollow ; enteron, intestine,) or animals with an alimentary canal communicating
with the general cavity of the body ; and, II. ECHINO-
DERMATA, (echinos, a spine ; derma, skin,) or spiny-
skinned animals. Other characteristics, however, besides
those signified in the names of these sub-types are necessary to be considered.

4. The CŒLENTERATA are radiate animals with a distinct body-cavity, whose walls consist of two layers of
cellular tissue, an outer (ectoderm) and inner, (endoderm,)
and contain nettling thread-cells, or small sacs full of
fluid connected with barbed filaments, capable of being
projected for stinging purposes. Most of these animals
have hollow tentacles round the mouth. There are two
large classes of Cœlenterates : I. The HYDROZOA, which
have no digestive cavity separate from the rest of the
mass which forms the body, and whose reproductive organs are external ; and, II. ACTINOZOA, which have a
digestive canal distinct from the rest of the body, suspended by radiating partitions, called mesenteries ; and
whose organs of reproduction are internal, placed on the
mesenteries. The first of these classes may be repre-

* Agassiz.

sented by the Hydra, and the latter by the Sea-anemone, or Actinia.

The *Hydra* is named after a fabled monster that reproduced its heads as fast as they were cut off. The genus comprises two species, the green and the brown Hydra, (*H. viridis* and *H. fusca.*) (Fig. 74.) They are minute creatures, about a quarter of an inch long, generally found on the under surface of aquatic plants, attached by a disk, while their long tentacles float downward in search of prey. The body is a simple tube, or cavity, and the tentacles are supplied with "lasso-threads," or nettling thread-cells. In the early summer small buds grow from the base of the body, which grow into the likeness of the par-

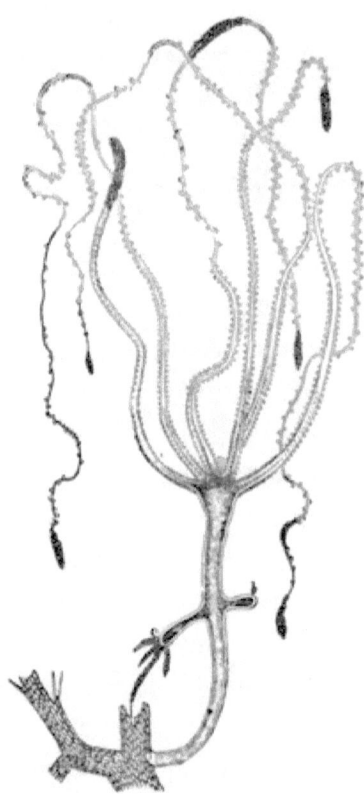

FIG. 74.—*Hydra fusca*, with a young bud at *b*, and a more advanced bud at *c*.

ent, and then are detached. Sometimes a second crop of buds arise from the first before it is separated. Later in the season eggs form from modified cells of the inner layer, which burst through the outer layer, become free, and develop into new Hydræ.

These animals are nearly allied to the Protozoa, since the differentiation of function in the bioplasts is incomplete. Hence the wonderful powers of propagation in these creatures, which have astonished naturalists ever since Trembley first discovered them, in 1744. He says: "I have opened a polyp on my hand, extended it, and cut the simple skin of which it is formed in every direction; I have reduced it to little pieces, and, in a manner, minced it. These little pieces of skin, both those which did and those which did not possess arms, became perfect polyps." Many curious multiple forms have been produced by experiments on these animals. By slitting the body into two branches, and these branches again into others, a tree-like form may be produced, each branch giving rise to a distinct head and tentacles. Or one may be turned inside out like a glove, so that the outer skin becomes the lining of the stomach-cavity, with a transposition of the functions of each.

Order 1. *Hydroida.* This order is composed of animals built on the pattern of the Hydra, just described. They are either single, as the Hydra, or compound. The latter are subdivided into the three families of Campanularians, Sertularians, and Tubularians. They are grouped in clusters or colonies on a common axis or stalk, (*cœnosarc.*) Each hydra-like organism is called a *polypite.* New polypites arise as outgrowths from the common stem of the colony, so that the stomach of each is continuous with the tubular center of the stalk, producing a community of nutrition in the colony. In Chap. III., Sec. 14, reference was made to the alternation of generations which this order of animals so strikingly illustrates. This process in the life-history of the Hydroids is briefly as follows: The Polyp, a fixed animal, increases for

15*

awhile by budding, but at a certain period gives birth by subdivision, to free swimming Medusæ, or Jelly-fish Each of these, after pursuing for a time its own course of life and development, produces eggs which change into ciliated bodies (*Planula*) similar to some of the Infusoria. After a while each of these becomes stationary, fixes itself to some weed or rock, and becomes a polyp, or Hydroid.

Those Medusæ which swim by the contraction of their umbrella-like disk were formerly called *Pulmogrades;* those which swim by vibratile cilia attached to arms, *Ciliogrades;* those which float by an expansive bladder, *Physogrades;* and those furnished with arms, or cirri, *Cirrigrades.* Another classification divided them into " naked-eyed " and " hidden-eyed " Medusæ. Since more thorough research has shown their relation to the Hydroids, the Medusæ have been considered in reference to the families of Hydroids from which they spring.

The *Tubularian* family (*tubulus*, a little tube) consist of Hydroids, sometimes simple, but generally compound, united by a common trunk or cœnosarc, which has an external horny coat, or *polypary*. Sometimes the tube is jointed with the tentacles placed in a whorl round each joint, (*Tubularidæ divisa*,) sometimes it is undivided, (*T. indivisa.*) Sometimes the polypary is much branched, (as in *Eudendrium*,) but in the majority it is not branched. A few species have no hard polypary, (as *Corymorpha nutans*,) but simply a white fleshy stem. The polyps of this family have no protecting cups. The Medusæ bud from the stem.

The *Sertularian* family (*Sertula*, a little wreath) is generally regarded as a sea-weed by sea-side visitors, but a very cursory examination with a pocket lens will suffice to show the horny and branched polypary, with its little cups, (*hydrothecæ*,) which contain and protect the polypites. In some of the Sertularians the Medusæ wither on the stock, never becoming free.

The *Campanularian* family (*Campanula*, a little bell) resemble Sertularians, except that the cups (*hydrothecæ*) are stalked and terminal instead of being lateral and sessile, as in the latter. The reproductive calyces, or ovarian capsules, may contain many Medusæ buds developed one below the other, which are set free by the bursting of the cell. (Fig. 75.)

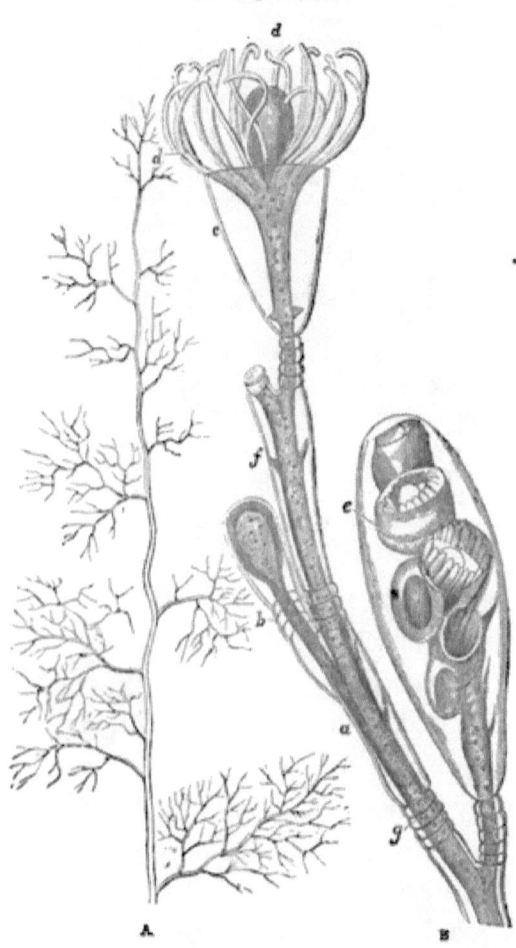

FIG. 75.—Campanularia gelatinosa :—A. Upper part of the stem and branches, of the natural size. B. A small portion enlarged, showing the structure of the animal. *a.* Terminal branch bearing polyps. *b.* Polyp bud partially developed. *c.* Horny cell, containing the expanded polyp, *d.* *e.* Ovarian capsule, containing medusiform gemmæ in various stages of development. *f.* Fleshy substance extending through the stem and branches, and connecting the different polyp-cells and ovarian capsules. *g.* Annular constrictions at the base of the branches.

The ordinary Jelly-fish (*Medusa*, or *Acaleph*) is soft, gelatinous, and bell-shaped, with tubes radiating from

center to circumference, where they connect with a circular canal. The margin is fringed with stinging tentacles. The radiating parts are in multiples of four. These gelatinous bells, varying from the size of a pea to a foot or more in diameter, float, mouth downward, in the sea, and propel themselves by flapping

FIG. 76. — Development of Sarsia. 1. Polyps described as Syncoryne, natural size. 2. A polyp, magnified. *a*. Polyp stem. *b. c. d. e.* Medusoid buds, in various stages. *f.* Tentacles of the polyp. 3. Free Medusa of the genus *Sarsia*.

their sides. (Fig. 76.) There are two representative forms of Medusæ, the *Lucernaria*, or Umbrella-acaleph, attached by a short pedicle, and having tentacles disposed in eight groups around the margin, and not less than eight radiating canals; and *Discophora*, the ordinary Jelly-fish, free and oceanic, with four radiating canals in the disk, which ramify and open into a circular canal around the mouth of the disk.

Order 2. *Siphonophora*, or floating Hydroids, (*siphon*, a curved tube, and *phero*, to bear,) are free swimming or compound floating Hydroids, with an unbranched, or slightly branched, but muscular cœnosarc. The common stem of these colonies swims by means of enlarged and altered polyphites, whose stomachs are undeveloped and whose bodies are dilated. Some possess, also, a sac filled with air, which acts as a float, as the *Physalia*, (*physa*, a bubble,) or Portuguese Man-of-war, whose purple-crested air-sac and long tentacles attract such attention in tropical seas, and whose thread-cells inflict such painful stings when grasped by an incautious hand. (Fig. 77.) The *Porpita* (*porpe*, the ring of a shield) possesses an internal skeleton, or flat plate, of cartilaginous texture, which is cellular and lighter than water. Its lower surface contains a beautiful fringe of blue tentacles, or cirri. In the *Velella* (*velella*, a little sail) a second

FIG. 77.—Physalia.

cartilaginous plate rises nearly at right angles from the upper surface of the horizontal one, serving as a sail to waft the little mariner from place to place.

CLASS II. ACTINOZOA, (*actin*, a ray; *zoon*, an animal.) This class embraces the Sea-anemones, the Corals, and the Ctenophora, (*kteis*, a comb; *phero*, I bear,) or comb-bearing Medusæ. The digestive cavity is suspended in the body cavity, like a small bag within a larger one, by vertical partitions, some of which extend from the body-wall to the digestive sac, but others fall short of it. Upon these septa, or mesenteries, are the organs of reproduction. The ectoderm is more highly developed than in Hydrozoa, and both mesenteries and body-walls are supplied with distinct sets of muscles. Cilia are present on the digestive tube, producing a current both respiratory and circulatory.

The *Actiniæ*, or Sea-anemones, are the much-admired forms so often seen in the rock-pools around our shores, sometimes called animal flowers, attached to the rocks by a flat disk, expanding their petal-like tentacles in search of prey, and, when uncovered by the retreating tide, contracting into small round gelatinous masses.

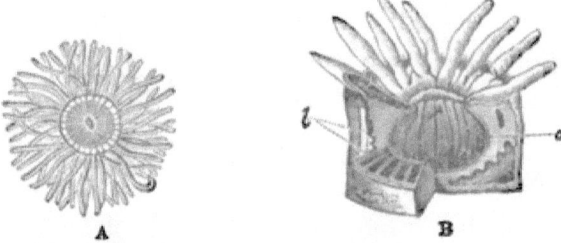

FIG. 78.—A. Sea-anemone, seen from above. B. Section of Sea-anemone. *a*. Cavity of stomach. *b*. Surrounding chambers.

The tentacles and partitions of the body are in multiples of six. Fig. 78 represents the internal form of Actinia.

The *Coral polyps* are Actinozoa, which secrete coral, generally composed of carbonate of lime, but it is occasionally horny, or a mixture of horny and calcareous matter. These polyps are usually found in colonies formed by a continuous process of budding. The compound mass is like a sheet of animal matter, fed and nourished by numerous mouths and many stomachs. Corals are of two kinds, the sclerobasic and sclerodermic corals. The polyps of the latter resemble Actiniæ in structure. The earthy matter is secreted between each pair of partitions, so that the skeleton of a single polyp (or *corallite*) is a short tube with vertical septa

FIG. 79.—Corals.

radiating toward the center. The *Fungia*, or Mushroom coral, is disk-shaped, and differs from others in not being either fixed or compound. It is simply the skeleton of a single polyp, showing a radiating secretion of calcareous septa. The various kinds of budding in compound coral-polyps give rise to a variety of shapes, either dome-like or

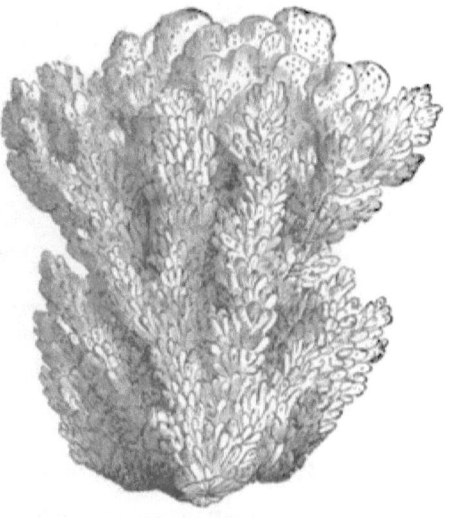

FIG. 80.—Madrepore.

branching. *Astræa* is a hemispherical mass covered with large cells. *Meandrina*, or "Brain-coral," has the mouths of the polyps opening into each other, forming furrows. (Fig. 79.) *Madrepore* branches, like a tree, with pointed extremities. (Fig. 80.)

Sclerobasic corals are those which secrete coral by the outer layer of the inverted ectoderm. Most of these are of the order *Alcyonaria*, whose polyps are characterized by primate or fringed tentacles in multiples of four, while the sclerodermic corals generally belong to the order Zoantharia, with polyps having simple tentacles in multiples of five or six. The characters of Alcyonarian polyps may be seen by placing in sea-water some of those large yellowish, gristly masses, sometimes cast up by the sea, known as "dead men's fingers." From the

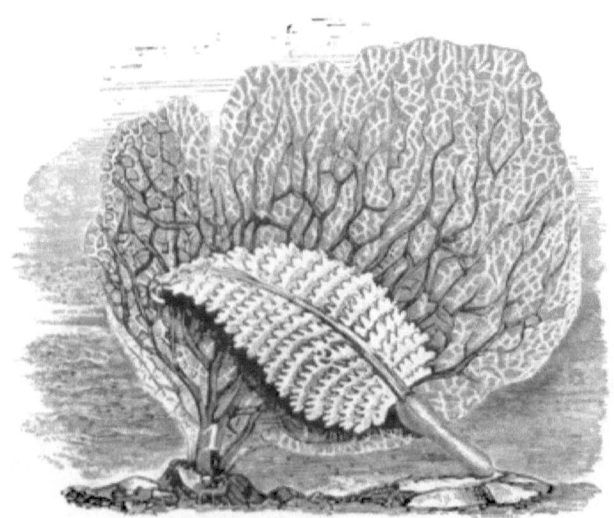

FIG. 81.—1. Sea-fan. 2. Sea-pen.

surface of each pore the tentacles round the mouth of the polyps will protrude, showing their general resemblance to Actinia. Minute spicules of calcareous matter are scattered throughout the mass. In *Gorgonia* such spicules, with horny matter, make up a continuous branching coral in the same plane, whose ramifications unite in a beautiful net-work. (Fig. 81.) In *Corallium rubrum*, the precious coral of commerce, the axis is of stony hardness, and branching like a shrub. In the

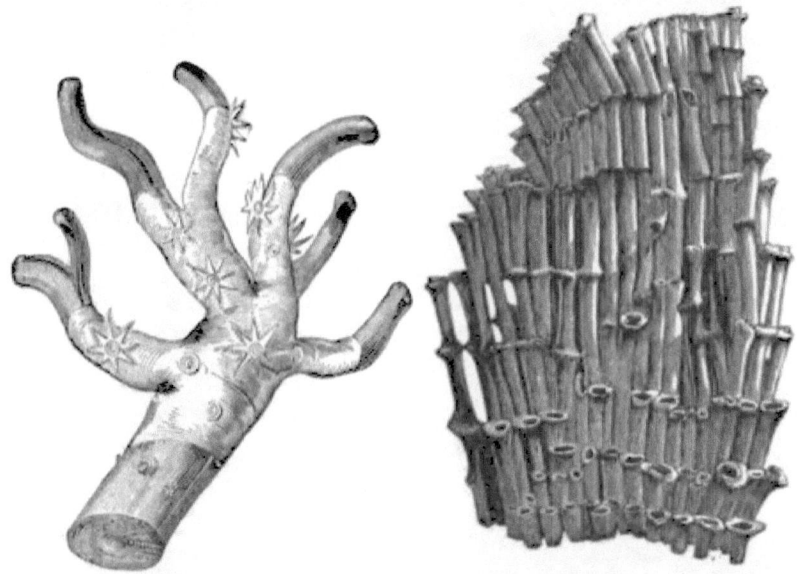

FIG. 82.—Red Coral.   FIG. 83.— Tubicora Musica.— Organ-pipe Coral.

living state the branches are covered with a red cœnosarc, (common flesh,) studded with polyps. (Fig. 82.) The feather-shaped sea-pens (*Pennatula*) have the extremities of their stems buried in sand. In some genera, as. *Virgularia*, the stem is prolonged to between three and four feet in length, while the polypiferous lobes are comparatively short. The red organ-pipe coral of the Indian Ocean, (Fig. 83,) with its table-like partitions and green polyps, belong also to this group.

16

The work of the reef-building polyps is extremely
interesting. They will not live in water whose mean
temperature is below 68° F., nor at a greater depth than
twenty fathoms, yet coral reefs are constantly found
which are several hundred fathoms thick. This apparent paradox is due to the fact that the land where coral
reefs are forming is constantly subsiding, and fresh living
corals are taking the place of the dead ones. If the
center of a reef sinks more quickly than the sides a
lagoon is left, surrounded by a circular reef of coral,
called an atoll; if an island rises in the middle of this
lagoon a barrier reef is said to be formed, (Fig. 84;)
while if the sea clearly intervenes between the reef and

FIG. 84.—A Coral Island.

the mainland, we have what is termed a "fringing reef."
Different species of polyps build these reefs. Madre-

pores, Millepores, and Gorgonidæ work chiefly at the top, next below we meet with Meandrinas, and lowest of all, with Astræans.

The *Ctenophoræ*, or comb-bearing Medusæ, exhibit traces of a nervous system in a ganglionic mass at the upper end, or pole, of the animal, with nervous filaments radiating to every part of the body. They are transparent gelatinous animals, which swim freely by means of bands of comb-like fringes or paddles. Their internal structure is quite complex, having a distinct alimentary canal, and ducts for the circulation of fluid. They are retained in the Radiate type, on account of the radiate arrangement of the bands of cilia and the presence of urticating organs on the tentacles, although their affinities would seem to place them elsewhere.

The *Beroë* and *Cydippe* (Fig. 85) and *Cestum Veneris*,

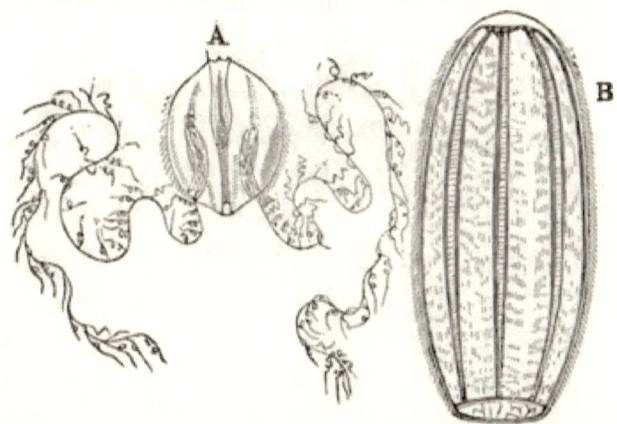

FIG. 85.—A. Cydippe pileus, with its tentacles extended. B. Beroë Forskalii, showing the tubular prolongations of the stomach.

or Girdle of Venus, belong to this order. In the latter, the sides are prolonged into a ribbon, although the mouth and digestive organs are confined to the middle of the body. In the day-time its waving cilia along the margins of the body glitter with the tints of the rainbow, and at night it appears like a long waving flame in the water.

5. The subtype of Radiate animals, called ECHINO-DERMATA, is distinguished by the possession of a nervous system, in the form of a pentagonal ring round the mouth; an alimentary canal, with oral and anal apertures; a peculiar system of circular and radiating canals; and a symmetrical arrangement of all the parts of the body around a central axis, in multiples of five. Some star-fishes (*Solaster*) have twelve rays. In all Echinoderms, probably, sea-water is freely admitted into the body-cavity around the viscera. The canals likewise contain water, which enters through a porous tubercle, the *madreporic plate*, or dorsal wart, best seen on the back of the star-fish and the sea-urchin. Some naturalists rank Echinoderms as Worms.

The *Crinoidea*, or Sea-lilies, so called from their resemblance to flowers, are fixed to the sea-bottom by a hollow, jointed, flexible stem, which carries the body, which is cup-shaped, with radiating arms, or tentacles. This order includes an immense number of fossil forms, but deep-sea dredging has brought up many living species, formerly thought to belong exclusively to the Mesozoic period. They all possess an internal skeleton of infiltrated calcareous matter, so that the entire animal consisted of thousands of stellate pieces, or joints, connected by animal matter. As each joint is furnished with at least two bundles of muscular fiber, one for extension and one for contraction, Dr. Carpenter estimates three hundred thousand such muscles in a single *Pentacrinus*—an amount of muscular apparatus far exceeding any that has been elsewhere observed in the animal creation. The family, COMATULIDÆ, or Hairstars—sometimes termed Feather-stars—in their young condition, resemble the *Encrinites*, or Sea-lilies, being supported on a long flexible stalk, composed of calcareous cylinders. At maturity they quit their attachment, and crawl about like other Star-fishes.

The order ASTEROIDEA, or Star-fishes, consists of animals with a flat central disk, having five or more arms, or lobes, radiating from it, and containing branches of the viscera. The skin is leathery, hardened by small calcareous plates, (eleven thousand or more,) but somewhat flexible. The mouth is below, and the rays are furrowed underneath and pierced with numerous holes through which pass sucker-like tentacles for locomotion and prehension. These furrows are named *ambulacra*, or avenues, from a fancied resemblance to a walk, or alley, in a garden. As the tentacles, or suckers, are only protruded from these spaces, they also have been called *ambulacra*. The arrangement for their protrusion will be described in connection with the Sea-urchins, as well as the *Pedicellariæ* (formerly believed to be parasitic organisms) found near the mouth.

About one hundred and fifty species of Star-fishes are known, divided into three groups: (1.) those having four rows of feet, as the common five-fingered Star-fish, or *Asterias;* (2.) those with two rows, as the many-rayed *Solaster*, or Sun-fish, and the pentagonal *Goniaster;* (3.) those with long slender arms, which are not prolon-

16*

gations of the body, and are not provided with suckers, as the *Ophiura*, or Brittle-star, (Fig. 86,) and *Astcrophy-*

FIG. 86.— Ophiura.

ton, or Basket-fish. The last group are inferior in structure, and resemble inverted stemless Crinoids. The digestive sac is confined to the disk, and the madreporic plate is underneath.

The order ECHINOIDEA, or Sea-urchins, contains those Echinoderms whose skin secretes calcareous plates, forming a hollow shell, covered with spines, and varying in shape from a sphere to a disk. The shell of an *Echinus* is made up of twenty rows, or zones, of plates, of which five pairs are ambulacral, pierced with minute pores for the protrusion of ambulacra, or sucker-feet, and five pairs alternating with the former are inter-ambulacral. The shell is developed from a membrane which lines the interior of the plates, and passes between the joints, so that additions can be made to their edges, by which means the shell grows and preserves the same relative proportions. The upper end of the shell, in addition to five small circularly disposed plates, carries five large genital plates. Each of these has a duct for the passage of ova or spermatazoa, and an *ocellus*, or eye-spot. One of these plates is the madreporic tubercle, with minute apertures communicating with the madreporic canal. Locomotion is effected by the hollow muscular feet, each of which communicates with a water sac; they also communicate with each other, so that as each sac contracts, water is forced into the corresponding tube, which is thereby elongated and protruded. (Fig. 87.)

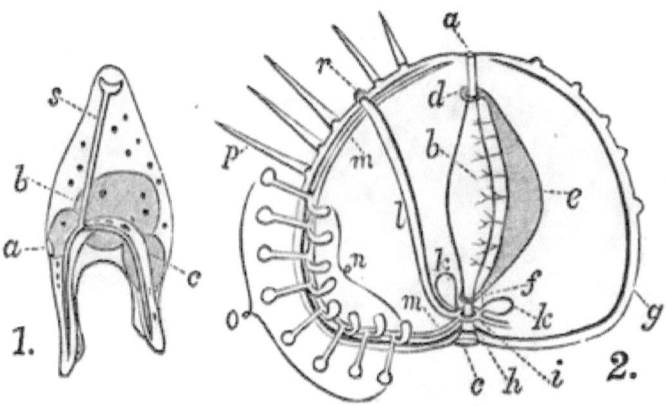

FIG. 87.—Morphology of Echinoidea. 1. Echinid larva. *a*. Mouth. *b*. Stomach. *c*. Intestine. *s*. Skeleton. 2. Diagram of Echinus. The spines and the ambulacra are represented over a small portion of the test; the vascular system is cross-shaded; the nervous system is represented by the black line. *a*. Anus. *b*. Stomach. *c*. Mouth. *d* and *f*. Vascular rings round the alimentary canal. *e*. Heart. *g*. Test. *h*. Nervous ring round the gullet. *i*. Ambulacral ring, or circular canal round the gullet. *k k*. Polian vesicles. *l*. Sand canal. *m m*. Radiating ambulacral canal. *n*. Secondary ambulacral vesicles. *o*. Ambulacra, or "tube-feet." *p*. Spines. *r*. Madreporiform tubercle.

The shell of the Echinus is covered with semiglobular warts, or beads, each of which during life supports a sculptured spine with a hollow at its base, forming with its muscles and ligaments a

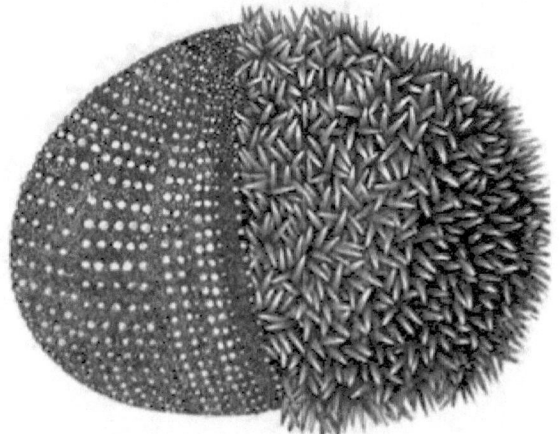

FIG. 88.—Shell of Echinus, or Sea-urchin; on the right side, covered with spines; on the left, the spines removed.

ball and socket joint, subsidiary to locomotion. (Figs. 88 and 89.) *Pedicellariæ* are minute, almost microscopic,

jointed spines, scattered all over the shell of the Echinus,
and terminated by a three-fold claw, capable of being

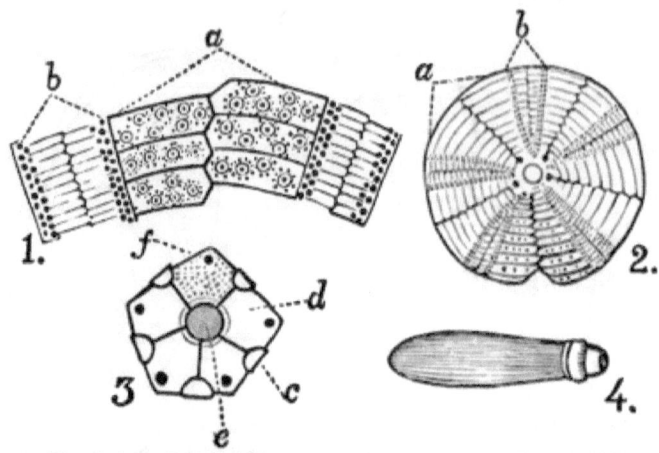

FIG. 89.—Morphology of Echinoidea. 1. Portion of the test of Galerites hemisphericus
enlarged, showing the inter-ambulacral area (*a*) and the ambulacral areas, (*b*.) 2. Galerites hemisphericus viewed from above. *a*. Inter-ambulacra. *b*. Ambulacra. 3. Genital
and ocular disk of Hemicidaris intermedia, enlarged. *c*. Ocular plate. *d*. Genital plate.
e. Anal aperture. *f*. Madreporiform tubercle. 4. Spine of the same. (After Forbes.)
The tubercles are mostly omitted on figs. 2 and 3 for the sake of clearness.

closed like a pair of forceps upon animalculæ or other
offensive matter that may tend to obstruct its shell. One
carries the rejected matter to another till the surface is
completely free.

The mouth of an Echinus contains the most complex
and perfect dental apparatus in all the Animal Kingdom,
although occurring in a type generally considered of a
low grade of structure. It sets at naught all theories of
Evolution, since in our progress from the simplest forms
of life it is the first instance of a dental apparatus, and
the most perfect of all. It is composed of five accurately-
fitting vertical pyramids, each provided with a rod-like
tooth, worked by a couple of beautifully arranged muscles. (Fig. 90.) The intestine is tortuous and connected

by delicate mesenteries to the shell. These animals possess a heart with an aorta surrounding the gullet and intestine. The blood is aërated by exposure to the oxygen mixed with the water which is constantly circulating over the vascular mesenteries.

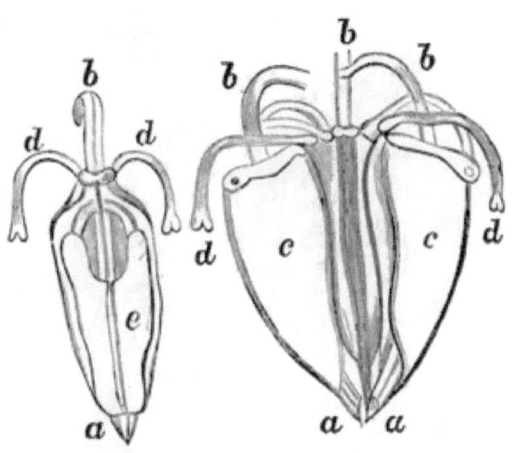

The metamorphosis of Echinus is very curious. The embryo is a free swimming minute ciliated creature, strangely like a painter's easel, and

FIG. 90.—Dentary Apparatus of Echinus, or Aristotle's Lantern. The right-hand diagram shows three of the teeth in position. *a a*. Cutting edges of the teeth, which are extremely hard. *b*. Fibrous roots of the teeth. *c c*. Opposed bony surfaces of the jaws. *d d*. Arched processes. The left-hand diagram shows an isolated pyramid. *e*. External surface. Other letters as before.

hence called a Pluteus. [*Pluteus*, a penthouse.] This passes through a strange cycle of changes. The digestive canal appears in the middle of the frame, which gradually disappears, the future Echinus is sketched in, and a radially symmetric animal results, totally unlike its predecessor. (Fig. 91.)

Regular Echini, as the common *Cidaris*, are nearly globular, and the oral and anal openings are opposite. Irregular Echini, as the *Clypeaster* and *Spatangus*, are flat, or discoid, with hair-like spines, and the rows of ambulacra form a five-rayed star on the back of the shell. *Spatangus* has no dental apparatus.

The order HOLOTHUROIDEA, embraces what are commonly known as Sea-slugs, Sea-cucumbers, or Trepangs. The body is elongated and soft, with a tough contractile skin containing calcareous spicules. One end, the head, has a simple aperture for a mouth, encircled with feathery tentacles. In the *Holothuriæ* proper, locomotion is effected by rows of ambulacral tube-feet, but in the *Synaptidæ* there are no ambulacra, and the animal moves by means of anchor-shaped spiculæ which are scattered in the integument. Animals of this order have the singular power of ejecting all their internal organs, surviving the loss of these parts, and afterward reproducing them. Their vermiform larva has no skeleton.

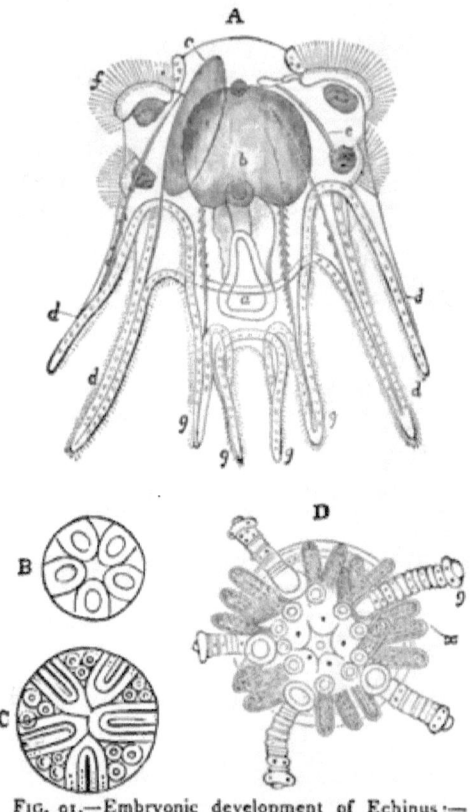

FIG. 91.—Embryonic development of Echinus:— A. Pluteus larva at the time of the first appearance of the disk. *a.* Mouth in the midst of the four-pronged proboscis. *b.* Stomach. *c.* Echinoid disk. *d d d d.* Four arms of the Pluteus body. *e.* Calcareous frame-work. *f.* Ciliated lobes. *g g g g.* Ciliated processes of the proboscis. B. Disk, with the first indication of the cirrhi. C. Disk, with the origin of the spines between the cirrhi. D. More advanced disk, with the cirrhi and spines projecting considerably from the surface. (N.B. In figs. B, C, and D, the Pluteus is not represented, its parts having undergone no change, save in becoming relatively smaller.)

6. The Radiate type of animal life well illustrates the intellectual plan, or typical design, of living forms, and

contains many instances totally unaccountable on any scheme of material gradation whatever. The nettling thread-cells, or *Cnidæ* in the Hydroids, the peculiar alternation of generations in the Medusæ, the great amount of muscular development in the Crinoida, the pedicellariæ, and the dental apparatus of Echinus, are all examples of structural arrangement for a purpose, and make against the theory of evolutional development, or survival of the fittest. Each of these structures are the most perfect of their kind, and seem to have no previous structure from which they have developed, as they have left no succeeding apparatus analogous or homologous to them.

CHAPTER XIII.

MOLLUSCA.

I have seen
A curious child applying to his ear
The convolutions of a smooth-lipped shell,
To which, in silence hushed, his very soul
Listened intensely, and his countenance soon
Brightened with joy ; for murmuring from within
Were heard sonorous cadences whereby,
To his belief, the monitor expressed
Mysterious union with its native sea.
—WORDSWORTH.

1. THE type of Mollusca, or soft-bodied animals, is indicated by the name, derived from the Latin *mollis*, soft. Like other types it embraces species of various degrees of complexity of structure, and of various forms. It includes soft-bodied, unjointed animals, possessing a muscular skin, or mantle, generally protected by a calcareous shell, and whose nervous system is scattered. It is subdivided into 1. MOLLUSCOIDA, containing the classes *Polyzoa, Tunicata*, and *Brachiopoda;* and 2. TRUE MOLLUSCA, embracing the classes *Lamellibranchiata, Gasteropoda*, and *Cephalopoda.*

2. POLYZOA (Gr. *polus*, many, and *zoon*, animal) derive their name from the fact of their living in clusters or colonies. They are sometimes called *Byozoa*, (Gr. *byon*, moss, and *zoon*, animal,) or Sea-moss. They greatly resemble the Hydroid polyps, but from the greater complexity and character of their organization they have been

removed to this type. The cells of a group are not connected with a common tube, as in Cœlenterates, and each
animal possesses a
distinct alimentary
canal and nervous
system. Sometimes
the colonies are foli-
aceous, or leaf-like,
as the Sea Mat,
Flustra, (Fig. 92,)
and at others plant-
like, as the *Plumat-*
ella. (Fig.93.) They

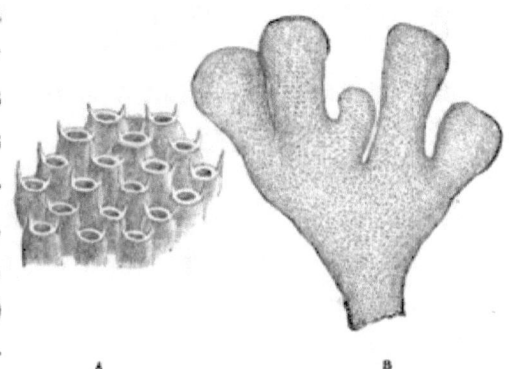

FIG. 92.—Sea Mat, (*Flustra foliacea*.) A. Magnified.
B. Natural size.

sometimes spread over rocks and sea-weeds like delicate
lace-work, and the majority are coral-making animals, or

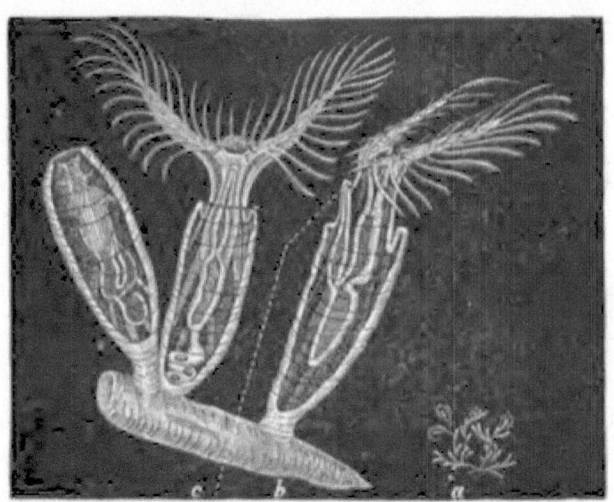

FIG. 93.—Plumatella. *a.* Natural size. *b.* A group enlarged. *c.* Anal orifice.

secrete carbonate of lime. The mouth of each animal is
surrounded by ciliated tentacles which serve for prehen-

17

sion, circulation, and respiration. Many species are furnished with organs of a remarkable and peculiar kind,
called *Avicularia*, (*avicula*, a little bird,) or "bird's
heads," which during life, and even after the death of
the animal, keep up a continual motion, see-sawing, and
snapping, and opening their jaws in the most singular
manner. Their use is unknown, but Mr. Gosse conject-
ures that they may seize and hold minute animals until
decomposition attracts a crowd of Infusoria, which may
serve the Polyzoan for food. Some species of Polyzoa
are found in fresh water.

3. TUNICATA, named from the Latin *tunica*, a cloak,
is a class of Molluscoida which are enveloped in a tough,
leathery sac, or "test." This sac is double-walled, but
not capable of protrusion. The mouth of the animal
opens into the bottom of a respiratory sac whose walls
are lined by a net-work of blood-vessels. The tubular
heart exhibits the curious phenomenon of reversing its
action at brief intervals, so that the blood oscillates
backward and forward in the same vessels. The wall of
the tunic contains cellulose, which is generally a vegetable product.

These bottle-shaped creatures are found in the ocean,
"solitary," attached to rocks or sea-weed, and often
glued together in bunches. Sometimes they are in "so-
cial" groups, as in Fig. 94, or "compound," as Fig. 95.

The *Salpæ* are free swimming, transparent *Ascidians*,
(*askos*, a bag; *cidos*, like,) or Tunicates, often found adhering to each other in long chains, which give birth to
solitary individuals of different form by alternation of
generations.

Young Tunicata swim, like tadpoles, by a tail, which contains a peculiar rod-like body, consisting of nucleated

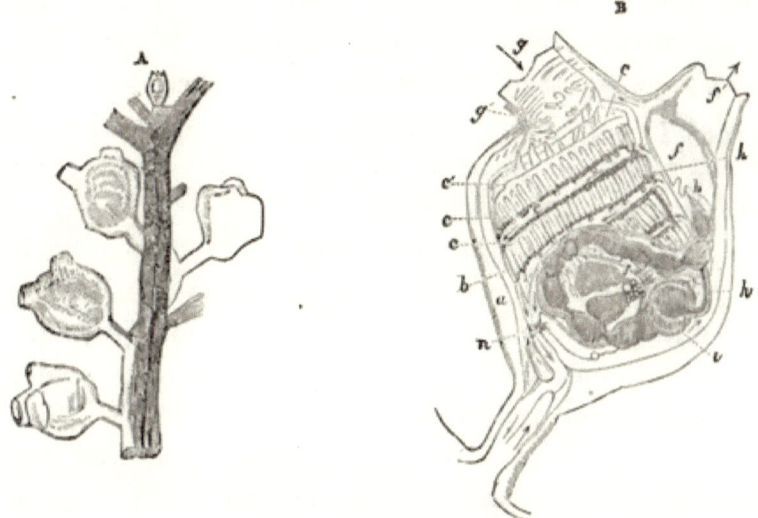

FIG. 94.—A. Group of Perophora, (enlarged,) growing from a common stalk :—B. Single Perophora. *a.* Test. *b.* Inner sac. *c.* Branchial sac, attached to the inner sac along the line *c'. c'.* *e. e.* Finger-like processes projecting inward. *f.* Cavity between test and internal coat. *f'.* Anal orifice or funnel. *g.* Oral orifice. *g'.* Oral tentacula. *h.* Downward stream of food. *h'.* Œsophagus. *i.* Stomach. *k.* Vent. *l.* Ovary. (?) *n.* Vessels connecting the circulation in the body with that in the stalk.

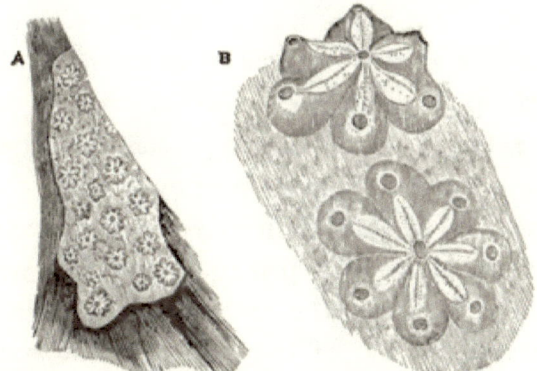

FIG. 95.—Botryllus violaceus : A. Cluster on the surface of a Fucus. B. Portion of the same enlarged.

cells like the *chorda dorsalis*, or *notochord;* an elongated mass of cells in the Vertebrate embryo, which is after-

ward replaced by the vertebral column. From this resemblance the partisans of evolution have claimed that
this simple cellular structure is the prototype of that
which distinguishes the higher animals, and that from
the simple Ascidian the Vertebrate has been developed.
Such foreshadowings of higher types is not uncommon.
It will require, however, much greater evidence to prove
transmutation than such resemblances.

4. BRACHIOPODA are protected by a bivalve shell,
which is lined by an expansion of the integument, or
"mantle." The valves of the shell are applied to the
dorsal and ventral sides of the body. The ventral valve
is usually larger and more convex than the other; but
they are symmetrical, that is, a vertical line from the
hinge divides the shell into equal parts. The ventral
valve generally has a hole, or *foramen*, through which a
fleshy foot protrudes for attachment. The mouth is
furnished with two long arms, fringed with cirri, generally coiled up and supported by a bony frame-work in
the shell—the "carriage-spring apparatus." As there
are no gills, the animal respires by the arms or the mantle. Brachiopods were once very abundant, over two
thousand extinct species having been described; but
less than one hundred species are now living.

In all the Molluscoida the nervous system consists of
a single ganglion, or of a principal pair with accessory
ganglia placed between the oral and anal apertures, or
on the ventral surface of the body. Some naturalists
connect them with the Worms.

5. LAMELLIBRANCHIATA (Lat., *lamella*, a plate; Gr.,
bragchia, gill) comprise the ordinary bivalves, as the

Oyster, Mussel, and Clam, and are characterized by the possession of lamellar gills. The shells differ from those of Brachiopods in being placed on the right and left sides of the body, so that the hinge is on the back of the animal, and in being generally unequilateral and equivalved. They are sometimes termed CONCHIFERA, or shell-fish, (Lat., *concha*, a shell ; *fero*, I carry.)

The shells of Mollusks are epidermal structures. The mantle, or loose skin, secretes calcareous matter in layers, converting the epidermis into shell. The microscopic structure is so characteristic that a thin section of a fragment often suffices to determine the group to which it belongs. A large class of shells is formed like the Oyster, of three parts; the external epidermis, brown and of a horny texture ; the prismatic portion, consisting of minute columns set perpendicularly to the surface ; and the internal nacreous, or pearly layer, made up of very thin plates whose edges overlap and form wavy lines. In many cases the prismatic and pearly layers are traversed by minute tubes. The pearls of commerce, found in the mantle of some Mollusks, are similar in structure to the shell ; but what is the innermost layer in the shell is outside and much finer in the pearl, which is formed around some nucleus, as an organic particle or grain of sand.

Shells of one piece are called " univalves," as the snail. Others, as the Clam, are of two parts, and are called " bivalves." The ribs, ridges, and spines on the outside mark successive periods of growth, and correspond with the age of the animal. Figs. 96 and 97 show the principal parts of ordinary bivalves and univalves. The valves

17*

of a bivalve are generally equal, except in Brachiopods and in the Oyster. The umbones, or beaks, are a little in front of the center, and turn toward the mouth of the animal. The valves are joined by a ligament near the

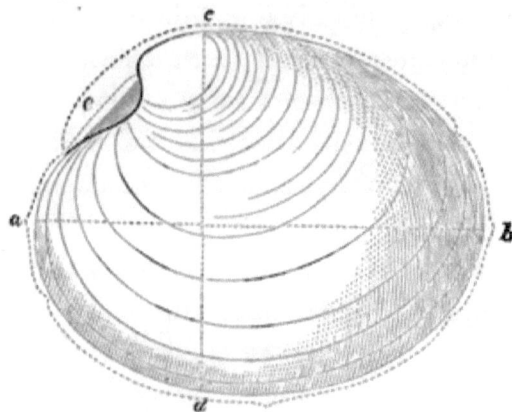

FIG. 96.—*a. b.* Length of the shell. *c. d.* Height. *e.* Lunula, above which is the summit. *d.* The ventral or inferior edge.

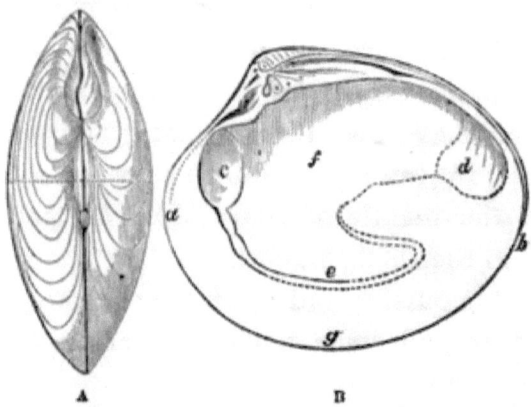

A   B

FIG. 97.—A. The line across marks the thickness of bivalves. B. *a.* Anterior extremity. *b.* Posterior. *c. d.* Muscular impressions. *e. f.* Palleal impression. *g.* Lower edge of the left valve.

umbones, and often also by a "hinge" formed by the "teeth" of one valve locking into cavities of the other. The aperture of a univalve is sometimes closed by a horny or calcareous plate, called an "*operculum.*"

Lamellibranchs breathe by four plate-like gills, two on each side, underneath the mantle. (Fig. 98.) In the higher forms the mantle is rolled up into two tubes, or siphons, for the inhalation and exhalation of water. The mouth opens into the stomach, which lies imbedded in a large liver, and the intestine, after a few turns, passes directly through the heart. (Fig. 99.) The nervous system consists of three pairs of ganglia, and the heart has two chambers, an auricle and ventricle, and, in some

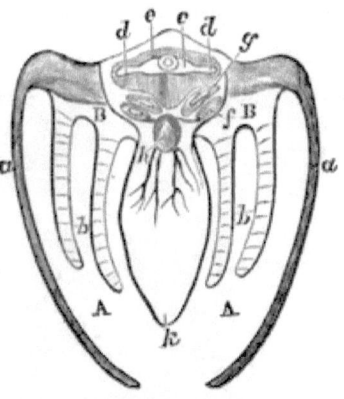

Fig. 98.—Diagrammatic Transverse Section of Anodon, through the heart. *a. a.* Lobes of mantle. *b. b.* Gills, showing transverse partitions. *c.* Ventricle of heart. *d. d.* Auricles. *e.* Pericardium. *f.* Glandular sac of organ of Bojanus. *g.* Vestibule, or middle sac. *h.* Venous sinus. *k.* Foot. A. A. Branchial or pallial chamber. B. B. Epibranchial chamber, communicating with cloaca.

cases, two auricles and a ventricle. The ventricle propels the blood into the arteries, by which it is distributed through the body. From the arteries it passes into the veins, and is conducted to the gills, where it is aërated, and is finally returned to the auricles.

A few Lamellibranchs are fixed, as the Salt-water Mussel, which hangs to the rocks by a cord of threads called "byssus," and the Oyster, which habitually lies on its left valve; but the rest have a foot by which they creep about. There are more than four thousand living species, fresh water and marine, which range from the line of shore to the depth of a thousand feet.

The muscular impressions on the shell, (*c. d.*, Fig. 97;) the presence of a pallial sinus, *e.*, which indicates the possession of siphons; the structure of the hinge, and the symmetry of the valves, are the chief characters for distinguishing genera and species of this class, which

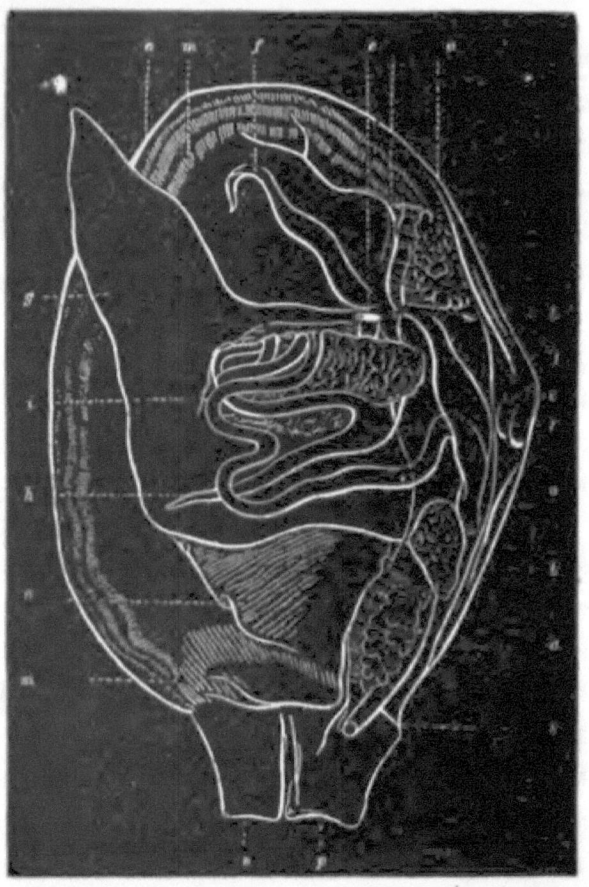

FIG. 99.—Anatomy of a bivalve Mollusk, (Mactra.) *a*. Shell-muscles. *b*. Ganglia. *c*. Heart. *d*. Liver. *e*. Mouth. *f*. Labial tentacles. *g*. Foot. *h*. Stomach. *i*. Intestine. *k*. Anus. *m*. Mantle. *n*. Branchiæ. *o*. Base of inhalent siphon. *p*. Base of exhalent siphon.

has been divided into groups, based on the possession or non-possession of siphons, as follows:

Section A. *Asiphonidæ*. Without respiratory siphons, so that the shell shows the pallial line simple, and not indented. As in the families of Oysters, (*Ostreidæ*,)

Mussels, (*Mytilidæ*,) Wing-shells, or Pearl Oysters, (*Aviculidæ*,) and River Mussels, (*Unionidæ*.)

Section B. *Siphonida.* Having siphons.

(1.) *Integro-pallialia.* Siphon short, pallial line simple, as in the families *Tridacnidæ*, **Cardiadæ**, (Cockles,) and *Cyprinidæ*, (Heart-cockles.)

(2.) *Sinu-pallialia.* Long siphons, pallial line sinuated, as in *Veneridæ*, (Clams,) *Mactridæ*, *Solenidæ*, (Razor-shells,) and *Pholadidæ*, (Boring-shells.)

6. GASTEROPODA, (Gr., *gaster*, stomach ; *pous*, foot.) This class derives its name from the fact that locomotion is usually effected by a muscular expansion of the under surface of the body, termed the "foot." It includes all the univalve shells, the naked slugs, the Dorsibranchs, the Pteropods, and the Multivalvular Chiton.

The body of most Gasteropods is unsymmetrical, the organs not being in pairs, but single, and on one side, instead of central. The mantle is continuous round the body, not bilobed, as in Lamellibranchs. A few, as the Garden-snail, have a lung, but the majority breathe by gills. The head is more or less distinct, and is provided with two tentacles, with auditory sacs, or rudimentary organs of hearing at their bases. The eyes are sometimes quite conspicuous. The Snail, for example, carries two *ocelli*, or simple eyes, on the tip of its long tentacles. Each consists of a globular lens, of short focus, which is a part of the transparent cornea, with a colored membrane (choroid) and a nervous net-work (retina) behind. The arrangement for retracting the eye and tentacle is seen in Fig. 100.

The mouth of Gasteropods possesses a peculiar strap-like organ, the *odontophore*, (*odous*, tooth; *phero*, I bear.) It is studded with three or more rows of lingual teeth, formed of silica, which are the serrated edges of minute plates, the number of which varies in different species; the garden Slug has one hundred rows with one hundred and eighty teeth in each row. (Fig. 101.) The strap, or " tongue," plays over a cartilaginous cushion, or pulley, connected with the lower jaw, and the teeth are renewed by fresh growths from the membrane beneath. The gullet is long, and frequently expands into a crop ; the stomach is often double, the anterior being a gizzard provided with teeth for mastication ; the intestine passes through the liver, and ends in the fore part of the body, usually on the right side. The heart is double, and a capillary system intervenes between the arteries and veins, but

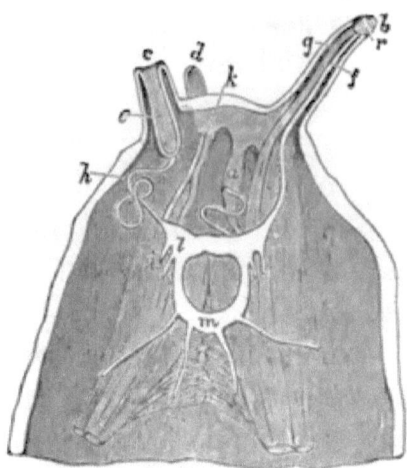

FIG. 100.—Head of a Snail bisected, showing structure of tentacles: *a*. Right inferior tentacle retracted within the body. *b*. Right superior tentacle fully protruded. *c*. Left superior tentacle partially inverted. *d*. Left inferior tentacle. *f*. Optic nerve. *g*. Retractor muscle. *h*. Optic nerve in loose folds. *i*. Retractor muscle of head. *k*. Nerve and muscle of left inferior tentacle. *l*, *m*. Nervous collar.

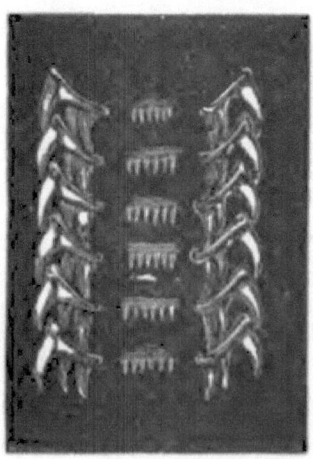

FIG. 101.—Palate of Buccinum undatum, as seen under polarized light.

the liver does not possess a distinct portal system, as in Vertebrates. (Fig. 102.)

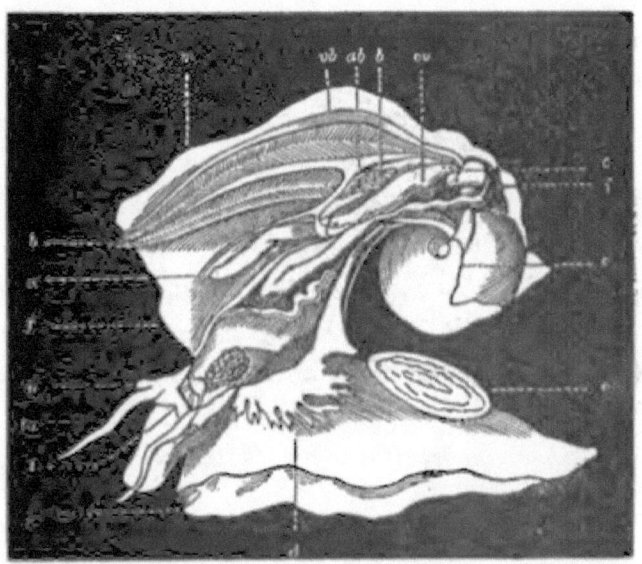

FIG. 102.—Anatomy of Turbo Pica: *p*. Foot. *o*. Operculum. *t*. Proboscis. *ta*. Tentacula. *y*. Eyes. *m*. Mantle opened longitudinally, to show the disposition of the respiratory cavity. *f*. Anterior border of the mantle, which, in its natural position, covers the back of the animal, leaving a wide slit by which the water enters the branchial cavity. *b*. Gills. *vb*. Branchial vein, returning to the heart, *c*. *ab*. Branchial artery. *a*. Anus. *i*. Intestine. *s*. Stomach and liver. *ov*. Oviduct. On the upper side of the neck are seen the cephalic ganglion, and the salivary glands; and at *d*. is shown a fringed membrane, which forms the lower border of the left side of the opening that leads to the respiratory cavities.

The univalve shell is generally a coiled tube, wound round a central axis, or *columella;* the nucleus, or earliest part of the shell being at the apex, and the portion last formed being the open mouth at the lower part, or base. The direction of the coil may be concentric, forming a discoidal shell, as *Planorbis*, but it is generally a true spiral. The mouth, or aperture, of the shell is entire in most vegetable-feeding Gasteropods, and notched or produced into a canal for the siphons in the carniv-

orous species. The former are generally land and fresh-
water forms, and the latter all marine.

Gasteropods comprise three fourths of all living Mollusks, and are representatives of the type.

Omitting a few rare forms, as *Dentalium* and *Carin-
aria*, we may divide the class into the following orders:

1. *Pteropods*, (Gr., *pteron*, wing; *pous*, foot,) which are
small marine floating Mollusks, whose main organs resemble a pair of fins or wings, whence the common name,
"Sea-butterflies." Many have a delicate, transparent
shell. The head is said to carry six appendages, armed
with several hundred thousand suckers, forming a prehensile apparatus unequaled in complication.

2. *Opisthobranchs*, (Gr., *opisthon*, behind; *bragchia*,

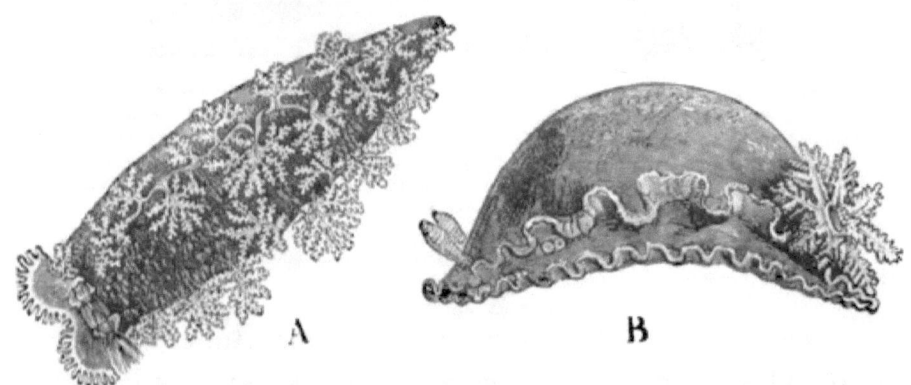

FIG. 103.—A. Tritonia Hombergi. B. Horned Doris.

gills.) These are generally naked Sea-slugs, a few only
having a small shell. The feathery gills are behind the
heart, (whence the name.) They are found in all seas,
generally on rocky coasts. When disturbed, most of
them draw themselves up into a lump of jelly or tough
skin. These naked-gilled Mollusks (*Nudibranchiata*) ex-

hibit a great diversity of form and a variety of beautiful colors. The Sea-lemon, (*Doris*,) the beautiful Tritonia, (Fig. 103,) the painted *Eolis*, the Sea-hare, (*Aplysia*,) which emits a violet or reddish fluid from the man-

tle when alarmed, and the Bubble-shell (*Bulla*) are examples.

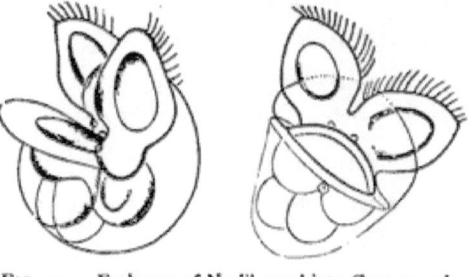

The embryo of the naked-gilled Mollusks is very minute, and resembles a Rotifer rather than a Mollusk.

FIG. 104.—Embryos of Nudibranchiate Gasteropods.

It is inclosed in a transparent nautilus-like shell, provided with an operculum. (Fig. 104.)

FIG. 105.—Snails and Slugs.

3. *Pulmonates* (having lungs) are air-breathing Gasteropods, represented by the common Snail. They have

the simplest form of lung—a cavity lined with a delicate net-work of blood-vessels, which opens externally on the right side of the neck. This opening is covered by a valve. They are found in all zones, but most where lime and moisture abound. All feed on vegetable matter. A few are naked, as the Slug; some are terrestrial; others live in fresh water. The Land-snails, as the *Helix*, *Bulimus*, and *Limax*, (Slug,) have four horns, the short front pair being the true tentacles, and the long hinder pair the telescopic eyes. The Pond-snails, as *Limnæa* and *Planorbis*, have no eye-stalks, the eyes being at the base of the tentacles. They are obliged to come to the surface of the water to breathe. (Fig. 105.)

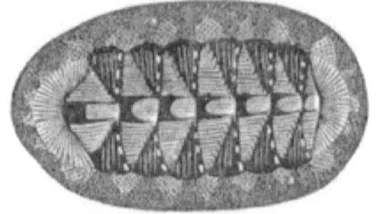 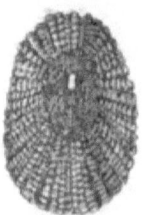 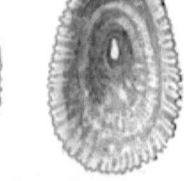

FIG. 106.—Chiton.   FIG. 107.—Fissurella Reticulata.

4. *Prosobranchs.* Having gills in front of the heart. (Gr., *proson*, before; *bragchia*, a gill.) These are aquatic and generally marine animals, the most highly organized and most abundant of all the Gasteropods.

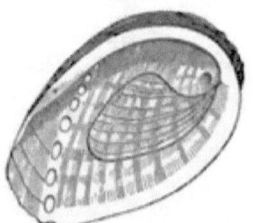

FIG. 108.—Ear-shell, or Haliotis. Reduced.

Among the lower forms are the singular *Chiton*, (Fig. 106,) covered with eight shelly plates; Limpet, (*Patella*,) well known to every sea-side visitor; and the beautiful Ear-shell Abalone, (*Haliotis*,) (Fig. 108,) often used for ornamental work and jewelry.

In the higher Prosobranchs the gills are comb-shaped and the sexes are distinct. The group includes all the spiral univalve sea-shells and a few fresh-water shells. Many have the aperture entire, as the fresh-water *Paludina*, the pyramidal *Trochus*, pearly *Turbo*, and

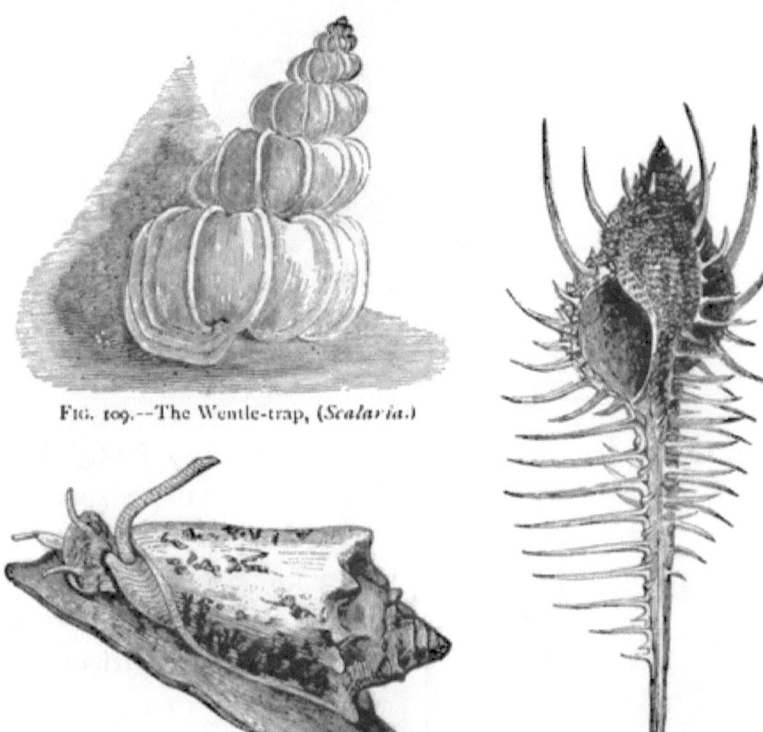

FIG. 109.--The Wentle-trap, (*Scalaria.*)

FIG. 110.--Volute Crawling.

FIG. 111.--Murex.

common Periwinkle (*Littorina*) from the sea. Others, the highest of the race, have the margin of the aperture notched or produced into a canal, and are carnivorous and marine ; such are nearly all the more beautiful sea-shells, as the Cowry (*Cyprœa*) Volute, (Fig. 110,)

Olive, Cone, Harp, Murex, (Fig. 111,) Whelk, (Fig. 112,) and Winged-shell, (Fig. 113.)

FIG. 112.—The Whelk, (*Buccinum*,) showing its operculum.

7. CEPHALOPODA, (Gr., *cephale*, head ; *pous*, foot.) The class of Cephalopods stands at the head of the Molluscan type. Some of its forms surpass in complexity of structure the highest Articulates, although not so representative of their type as the Gasteropods. They are aquatic free-swimming or creeping Mollusks, inclosed in a muscular mantle, and in some species having a univalve

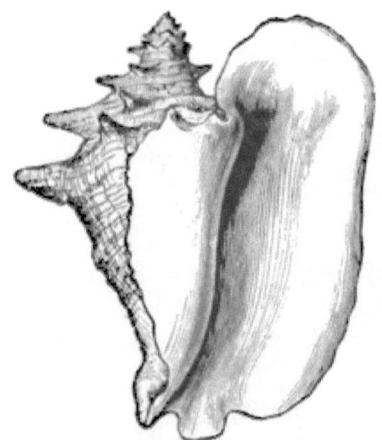

FIG. 113.—Strombus gigas, or "Winged-shell;" one fifth natural size. West Indies.

shell. The foot is divided into eight or ten long, waving, but strong tentacles, bearing numerous suckers, or *acetabula.* The adhesion of these suckers is so great that it is easier to tear away a limb than to detach it. Their mechanism may be understood from Fig. 114. The mouth has a horny beak, like a parrot's bill, but the jaws do not move vertically, like the bird's. A long gullet ends in a muscular gizzard, resembling that of a fowl. Below this is a cavity, the stomach or duodenum,

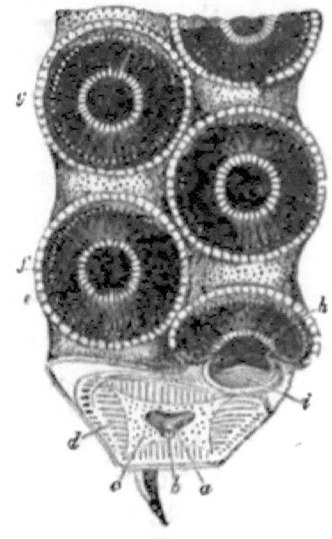

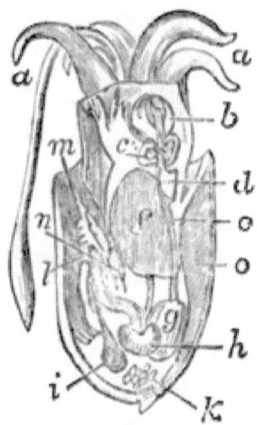

FIG. 114.—Suckers on the Tentacles of a Cuttle-fish : *a.* Hollow axis of the arm, containing nerve and artery. *c.* Cellular tissue. *d.* Radiating fibers. *h.* Raised margin of the disk around the aperture *f, g,* which contains a retractile membrane, or " piston," *i.*

FIG. 115. — Morphology of Cephalopoda. Sepia officinalis, laid open to show viscera, etc. *a.* Foot. *b.* Horny jaws. *c.* Principal ganglion. *d.* Salivary gland. *e.* Œsophagus. *f.* Liver. *g.* Stomach. *h.* Pyloric cæcum. *i.* Ink bag. *k.* Ovary. *l.* Aperture of atrial system. *m.* Branchiæ. *n.* Oviduct. *o.* Cuttle-bone.

which receives the bile from a large liver. The intestine is a tube of uniform size, which, after one or two slight curves, bends up, and opens into the " funnel " near the mouth. (Fig. 115.) The head is set off from the body by a slight constriction, and is furnished with a pair of large, staring eyes, which are constructed like

18*

the eyes of Vertebrates, except that there is no aqueous humor, and the lens, which is double, is bathed freely by the water in which the animals swim. The nervous system is more concentrated than in other Invertebrates : the cerebral ganglia are even inclosed in a cartilaginous cranium. All the five senses are present. The integument contains pigment sacs, or *chromatophores*, which sometimes tint the animal with variegated colors. It is probable that they in some way subserve the sense of sight, as the animal swims with its head backward. Some Cephalopods have an internal shell, secreted by a fold of the mantle, called the "cuttle-bone" or "pen."

Two or four pairs of plume-like gills are situated in the pallial cavity, into which the sea-water is admitted at one end and expelled through the funnel at the other by muscular contraction. These contractions serve both for respiration and locomotion, the pressure of the expelled water driving the animal in an opposite direction. The systemic heart pumps the blood all over the body, which then returns through capillaries into veins which conduct the blood back to the gills, where it is purified, and whence it is propelled to the heart by contractile sacs, called branchial hearts, placed at the base of each gill. In addition to other viscera, a large secreting sac, the *ink-bag*, is often present, containing a dark fluid which the animal ejects at will through a duct opening at the base of the funnel. The sexes are always distinct. During reproduction the spermatozoa are temporarily transferred to one of the arms, which becomes curiously altered and unfit for locomotion ; in this condition it is said to be *hectocotylized*.

1.) *Tetrabranchs.* This order has four gills, forty or more short tentacles, and an external chambered shell. The partitions of the shell are united by a tube, called a siphuncle, and the animal lives in the last and largest chamber. These chambered shells were once very abundant. More than two thousand fossil species are known, among which are the Nautilus, Ammonite, and Orthoceros. They have but one living representative—the Pearly Nautilus. This straggler of a mighty race dwells at the bottom of the Indian Ocean. The shell is well known, but only two or three specimens of the animal have been obtained.

2.) *Dibranchs.* Those having two gills. They are the most active of Mollusks, and the tyrants of the lower

FIG. 116.—The Paper Nautilus, (*Argonauta Argo.*) Fig. 1. Swimming toward the point *a.* 2. Walking on the bottom. 3. Contracted within its shell, which is partly embraced by the arms.

tribes. There are Cuttle-fish and Poulps (or Devil-fish) so large as even to be dangerous to a man who might be swimming near them, and the stories of novelists like

Victor Hugo have some foundation in the large size and repulsive aspect of these creatures. They crawl with their arms on the bottom of the sea, head downward, and also swim backward or forward, usually with the back downward, by means of fins, or squirt themselves backward by forcing water through their funnels.

The Paper Nautilus (*Argonauta*) (Fig. 116) and the Poulp have eight arms. The Squid (*Loligo*) and Cuttle-fish (Fig. 117) have ten arms, the additional pair being

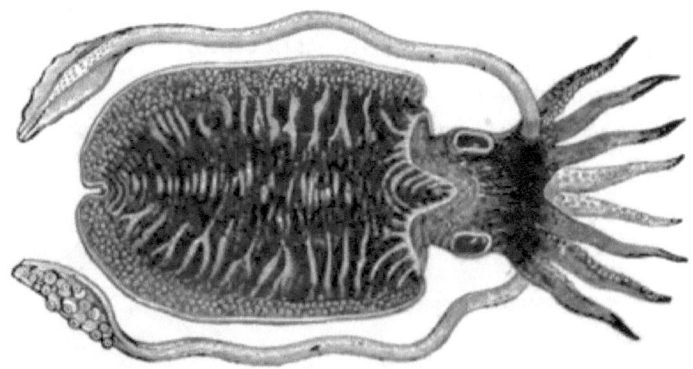

FIG. 117.—Cuttle-fish.

longer than the others. Their eyes are movable, while those of the Argonaut and Poulp are fixed. The Squid, used for bait by cod-fishermen, has an internal horny "pen," and the Cuttle has a spongy, calcareous bone.

CHAPTER XIV.

ARTICULATA.

"Yet wert thou once a worm—a thing that crept
    On the bare earth, then wrought a tomb, and slept !
    And such is man ; soon from his cell of clay
    To burst a seraph in the blaze of day !"—ROGERS.

1. THE Articulated type of animals (Lat., *articulus*, a joint) includes all which possess a distinctly jointed body, as Worms, Crustacea, and Insects. It contains a greater number and variety of forms than all the other types put together. The nervous system consists chiefly of a double chain of ganglia along the ventral surface of the abdomen, connected together by nerve-filaments. The part representing the brain is in the form of a ring encircling the gullet. The circulatory apparatus is a tubular structure running along the back, and communicating with the body-cavity. The limbs, when present, are jointed and hollow, and on the same side as the nerve-cords.

There are five classes of Articulates : the aquatic Worms and Crustaceans, and the air-breathing Spiders, Myriapods, and Insects. It must be remembered, in accordance with the principles so often referred to in the present work, that the order of classes in a type is one of relation rather than of structural rank. Classes cannot be arranged serially, any more than species, as if one was an improvement on another, by progressive development. In many respects Myriapods are like Worms, yet their heads show a resemblance to Insects. Some Spiders are less complicate than Myriapods, yet for their wonderful instincts Owen places them above Insects. Insects begin life as worm-like embryos. Classes in the articulate type depend on the equal or unequal development of the body-segments, and the number and form of appendages. Articulates with jointed appendages articulated to the body are called *Arthropoda*, (Gr., *arthron*, a joint; *podes*, feet.)

2. The class of WORMS, called, also, *Annelida*, or *Annulata*, (*Annulus*, a little ring,) includes animals with a soft skin and a body formed of a succession of rings, or movable joints. They differ from the Anthropoda in not having jointed limbs. A water-vascular system exists, but it has no connection with locomotion. The blood is often reddish, but the color does not depend on colored corpuscles, as in vertebrates. The circulatory apparatus is more highly developed than in Insects.

Some worms can only live as parasites upon the blood or juices of other animals, and in these the circulatory, water-vascular, and digestive systems become rudimental, the nervous system is undeveloped, the body-cavity often vanishes, and the reproductive organs alone are fully represented.

Order 1. *Tæniada;* (*tænia*, a tape.) Tape-worms, so called from their length and flatness. They live chiefly in the digestive canal of higher animals. Three species are occasionally parasitic in man. The head, which is the true animal, is provided with hooks or suckers, by which it adheres to the mucous membrane of its host.

It feeds by imbibition, (*osmosis*,) there being no mouth or alimentary canal. The joints, or segments, are called *proglottides*, (singular, *proglottis*,) and are but successive growths containing ova. The life-history of these worms is a curious instance of alternation of generations. The fertilized ova are set free by the decomposition of the joint, or proglottis. They are then swallowed by some animal, and the tough capsule is dissolved, setting free the embryo, which travels through the tissues of its host as a little oval body, bearing weak, hook-like, or boring spines. On reaching a suitable site, as the liver, it anchors, and the body dilates into a cyst, or sac full of water, (*Cysticercus.*) Many animals, formerly known as cystic worms, have been found to be but transitional stages of Tæniæ. In this condition the animal may

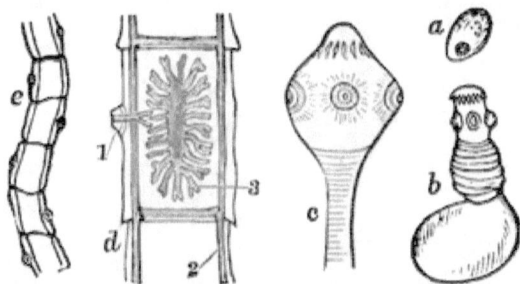

Fig. 118.—Morphology of Tæniada. *a.* Ovum with contained embryo. *b.* Cysticercus longicollis. *c.* Head of Tænia solium, (enlarged ;) the circlet of hooklets is at the top, and below them are those of the cephalic suckers. *d.* A single segment or proglottis magnified. 1. Generating pore. 2. Water vessels. 3. Dentritic ovary. *e.* Portion of Tape-worm, natural size, showing the alternating arrangement of the generative pores.

remain a long while and generate new cysts by budding, but when the flesh containing the "scolex," or resting-larva, is eaten by some other animal, the outer wall of the cyst dissolves, and becomes a true Tape-worm. The human Tape-worm has its cystic stage in "measly" pork, while the Tape-worm of the dog develops from cysts found in the hare, and that of the cat from cysts in the mouse; most cases requiring two animals as hosts for perfecting the growth of the worm. (Fig. 118.)

Order 2. *Trematoda ;* the Flukes. (Gr., *trema*, a hole.) These are flat or roundish parasitic worms. The intestine is branched, and, as in Cœlenterata, there is but a single opening, which serves for both mouth and anus. There are suckers at the anterior end of the disk. They are met with sometimes in the liver of the sheep.

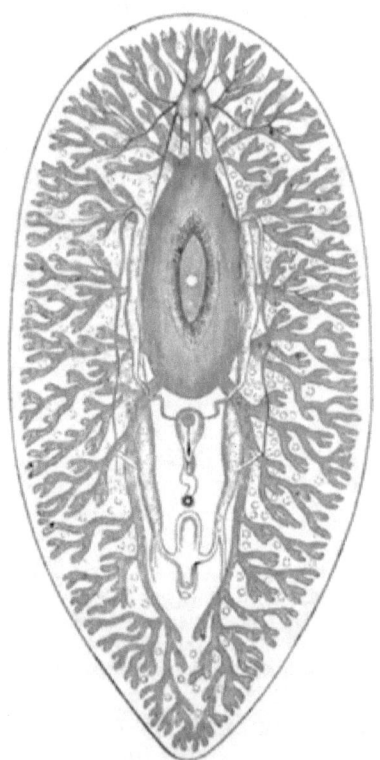

·Order 3. *Turbellaria.* These are non-parasitic, and may be found on the sea-shore, under stones, or in fresh-water pools, or on moist ground. They are small, ciliated, and flat worms, which glide with a slug-like motion over wet surfaces, or swim by the vibrations of their cilia. In the small flat *Planarians* the digestive cavity is greatly branched. (Fig 119.) In others it is a simple pouch, with no excretory orifice. In the larger forms it is elongated. Some

FIG 119.—Structure of Polycelis levigatus, (Planarian worm.)

of the largest (the *Nemerteans*) are like long ribbons; sometimes, as in *Borlasia*, being twelve feet long.

Order 4. *Acanthocephala;* (*akantha,* a thorn; *cephale,* head,) are rounded, parasitic worms, having a protrusible proboscis, armed with recurved hooks. Their structure is not unfrequently as simple as the Protozoa, having no alimentary canal whatever, and subsisting by absorption. Like the Tape-worms, they develop through an alternation of generations.

Order 5. *Gordiaceæ.* The horse-hair-like worm found in rain pools is an example of this order. It begins life as a little larva in mud or water pools. By means of its boring spines it pierces the body of a grasshopper, beetle, or other insect, where it becomes encysted; and grows often ten times as long as its host, when it becomes free and aquatic, and produces its eggs. Some of these, as the *Mermis albicans,* multiply so rapidly as to give rise to a popular belief that they fall as "worm-rains." They have remarkable tenacity of life, as they can be dried into brittle threads, and yet become active on being moistened.

Order 6. *Nematoidea,* (*nema,* thread; *eidos,* form.) Thread-worms, or round worms. These are both free and parasitic. Some of them, as the *Ascaris lumbricoides,* or common round worm, often infests the small intestines of children, while the *Trichina spiralis,* a minute worm found encysted in the flesh of swine, when introduced into the human body, multiplies so rapidly in the muscles as to give rise to dangerous, and even fatal symptoms. The "eels" in vinegar and sour paste also belong to this order.

Order 7. *Rotifera,* or *Wheel Animalcules.* These are microscopic in size, but so transparent that the details
19

of organization can easily be seen. The male rotifers are few and small, and have no digestive canal, but the females have a complete nutritive system, and many species are provided with an organ for mastication resembling an anvil acted on by two hammers, another instance of peculiarity of structure for a special end. These animals are capable of reviving on being moistened, after having been dried up, and that many times in succession. (Fig. 120.)

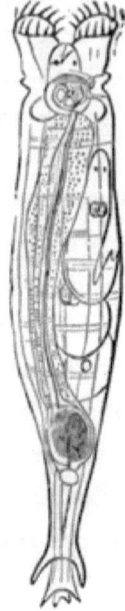

FIG. 120.—Rotifer vulgaris.

Order 8. *Gephyrea*, (*gephura*, a bridge,) so called in allusion to the apparent connection which they exhibit between Echinoderms and Articulates. They are sometimes called Spoon-worms, Squirt-worms, and Siphon-worms, (*Sipunculus*.) They have all the aspect of worms, but the circle of tentacles round the mouth show their affinity to Holothurians. They live in the sand, or seek protection in some empty univalve shell. Their elongated bodies contain a long, tortuous intestine, ciliated inside and outside. They have no locomotive processes, nor are there calcareous or silicious spicules in their skin. The mouth has a long proboscis.

Order 9. *Suctoria*, or Leeches. These are aquatic worms, with a soft, segmented body, provided with a suctorial disk at one or both ends. The mouth of the common Leech (*Hirudo medicinalis*) is armed with three horny, semi-lunar plates, with finely serrated teeth, which act as saws, enabling the leech to make incisions in the skin of its host through which to suck the blood.

Order 10. *Chætopoda*, or *Bristle-footed* worms. Some of these occur under the stones of the sea-shore, as the lug-bait of fishermen. (Fig. 121.) Others secrete a glutinous material from the surface, which cements sand and other foreign bodies into a tube. Others secrete calcareous matter, which forms a tubular residence, as the common *Serpula*, whose white, snake-like concretions abound on the stones and shells of the shore, and the *Spirorbis*, whose minute whorled shells dot the surface of many sea-weeds. Some of the *Nereids*, or Sea-centipedes, attain to a considerable size, one species being four feet long. The Sea-mouse (*Aphrodite*) also belongs to this order. The latter is clad with iridescent scales and bristles, or barbed spines. Those who bear the gills along the back have been called *Dorsibranchiates*. These gills are found close to the root of the dorsal oar, or bristle, and the blood is purified by being exposed to the oxygen held in solution in the sea-water. Those worms which live in tubes (*Tubicolæ*) have the gills developed only on the foremost segments of the body, and the dorsal and ventral oars of the other joints are rudimentary, but they have branching tentacle-like processes about the head. In *Serpula* one of the tentacles is formed into a lid, or operculum, with which the open mouth of the tube can be closed at will. (Fig. 122.)

FIG. 121. — Lob-worm, (*Arenicola piscatorum*,) a dorsibranchiate, showing the tufts of capillaries, or the external gills. The large head is without eyes or jaws.

The common Earth-worm (*Lumbricus*) has few and

small bristles, in the form of recurved hooks on each ring of the body, which assist in locomotion. It pos-

sesses no external gills, but respires by internal ciliated processes. The nervous system is often but little developed. The mouth is on the second segment, and the digestive canal is a straight tube, which is wide, and always full of earth, which these animals devour for the sake of the organic particles contained in it; the remaining part being cast out and heaped at the outlet of their burrows, as "worm-casts." For better division of the material swallowed the digestive canal

FIG. 122.—*Serpula.*

has a muscular gizzard about fifteen rings from the mouth. They are propagated by eggs.

3. The class of CRUSTACEA, (*crusta*, a crust or shell,) includes all Articulates with jointed legs and gills. They have a double, or complete circulation of blood; a dorsal tube, or heart, sending off a system of arteries, not found in insects; but the blood, as it leaves these tubes, escapes into the general cavity, as in other Articulates. (Fig. 123.) The shell, or *carapace*, has for its base a horny substance called *Chitine*. It is also found in the covering of Insects. In the Crab and Lobster there is a large proportion of carbonate of lime combined with this, rendering the carapace extremely hard. In others, a mixture of chitine and albumen gives rise to a softer integument. The rings of the body have considerable freedom of motion, by means of striated or voluntary

muscles. The normal number of joints is twenty-one, but two or three are often blended. To each of these

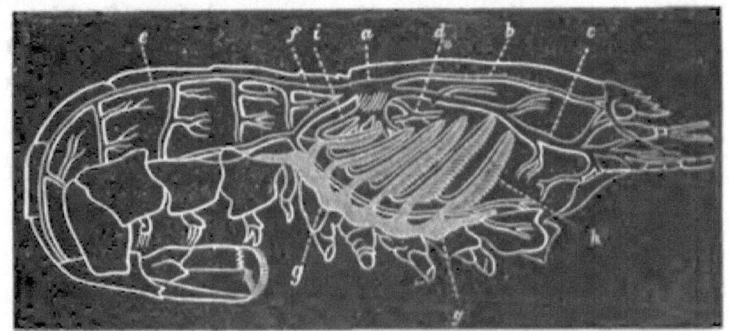

FIG. 123.—Circulating Apparatus of Lobster: *a.* Heart. *b. c.* Arteries to the eyes and antennæ. *d.* Hepatic artery. *e. f.* Arteries to thorax and abdomen. *gg.* Venous sinus. *h.* Gills. *i.* Branchial veins.

joints, except the last, there is attached a pair of members, the forms and uses of which vary in different species, and at different ages. These members are jointed, and covered with a similar envelope, or crust, to that of the body. As the body grows the carapace does not grow in the same proportion, rendering frequent moltings necessary. The entire covering is thrown off from body, feet, and antennæ in the most perfect manner. The Crustacea differ in habits as well as in structure. Most live in the water, but the *Land-crabs* inhabit the land. The Hermit-crabs (*Paguridæ*) live in the empty shells of Mollusks, which they seize, often killing the inhabitant. The majority of Crustaceans have jaws and organs of mastication, but some have no such organs, but live as parasites, especially on fishes, sucking their juices, and becoming strangely deteriorated. The alimentary canal in this class consists of a short gullet, a gizzard-like stomach, and a straight intestine. Crusta-

19*

ceans pass through a series of strange metamorphoses before reaching their adult form. The *Balanus*, or acorn-shell, which incrusts the rocks of the sea-coast in great numbers, begins life as an active, one-eyed free swimmer, called a "*Nauplius*," which after one or two molts becomes a pupa, inclosed in a bivalve shell by a folding of the dorsal portion. Finally it becomes a sed-

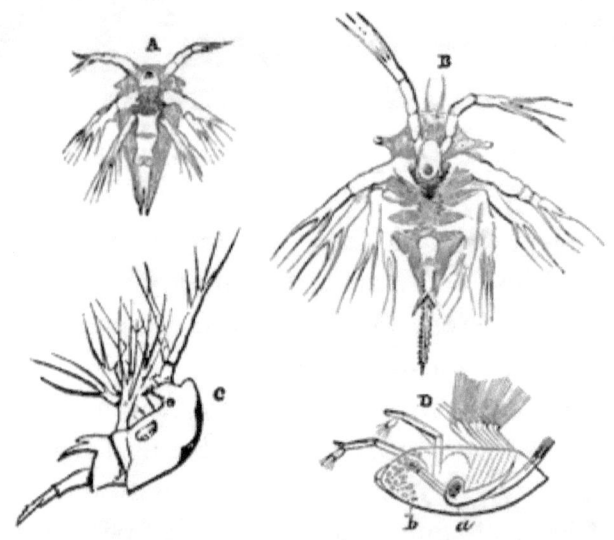

FIG. 124.—Development of Balanus balanoides: A. Earliest form. B. Larva after second molt. C. Side view of the same. D. Stage immediately preceding the loss of activity. *a*. Stomach.(?) *b*. Nucleus of future attachment. (?)

entary Cirripede, (*cirrus*, a curl ; *pes*, a foot.) (Fig. 124.) It will be convenient to divide Crustaceans into four groups, or orders.

1.) *Cirripeds*, distinguished by being fixed, having a shelly covering, and by their feathery arms. Such are Barnacles, (*Lepas*,) which have a peduncle, or stalk, and are often found on the backs of whales or on ship's bottoms, and Acorn-shells, (*Balanus*,) which are sessile.

2.) *Entomostracans,* which have a horny shell and no abdominal limbs ; represented by the little Water-fleas, (*Cyclops,*) (Fig. 125,) of our ponds, the King-crabs (*Limulus*) and the extinct Trilobites. The abdomen of the King-crab is reduced to a mere spine, the append-ages about the mouth are used for locomo-tion, and their eyes are smooth.

FIG. 125.—Water-fleas : 1. Cyclops communis. 2. Cypris unifasciata. 3. Daphnia pulex.

3.) *Tetradecapods,* small fourteen - footed species ; as the Wood-louse, or Sow-bug, (*Oniscus,*) found in damp places, and the Sand-flea, (*Gammarus,*) seen in summer on the sea-shore.

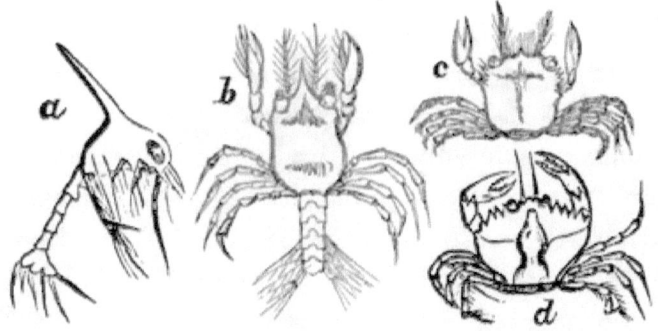

FIG. 126.—Metamorphosis of Crustacea, (*Carcinus mænas.*) *a.* Larval or first form. *b.* Second stage. *c.* Third stage. *d.* Final stage, in which the metamorphosis is complete.

4.) *Decapods,* having ten legs, as the Shrimp, (*Crangon,*) Cray-fish, Lobster, (*Astacus,*) and Crab, (*Cancer.*) Crabs differ from Lobsters chiefly in being formed for creeping at the bottom of the sea instead of for swimming, and in the abdomen, or tail, being a mere rudiment which folds into a groove under the enormous thorax. The curious metamorphosis of the Crabs is illustrated in Fig. 126. At first the embryo is a comical-looking animal, with a sort of spiked helmet on its head. It has two large eyes and a well-developed abdomen. It is called a "Zoea," and was formerly described as a distinct genus. After a succession of molts it becomes a perfect Crab.

4. ARACHNIDA (*arachne*, a spider) is a class much re-sembling the Crustaceans, having the body divided into a cephalo-thorax and abdomen. The head carries two, six, or eight eyes, which are not compound bundles of crystal rods covered by a common cornea, as in Crustaceans, but separate transparent cones surrounded with pigment. Antennæ are only two, if present, and are not used as " feelers," but serve prehension of food. Mandibles are always present, and in Scorpions the maxillary palps form pincers, or claws, like those of a Crab. Such claws are called *chelæ*, (*chele*, a claw.) Arachnids are all air-breathers, having spiracles which open into air-sacs, or tracheæ. The young of the higher forms undergo no metamorphosis after leaving the egg. The class is divided into three orders: Mites, Scorpions, and Spiders.

1.) *Mites* are the simplest forms of the class. They have no marked articulations, the head, body, and thorax being in one piece. They have no brain, but a single ganglion in the abdomen. They breathe by tracheæ. The mouth is formed for suction. Most are parasitic on animals or plants. Mites (*Acarus*) include the Cheese-mite, the Itch-insect, and many similar forms. The Ticks (*Ixodes*) have a piercing beak and a leathery skin.

2.) *Pedipalpi*, or Scorpions, have maxillary palpi ending in forceps, and a prolonged jointed abdomen. (Fig. 127.) Breathing takes place by pulmonary sacs, similar to spiders. The nervous and circulatory systems are highly organized. The last

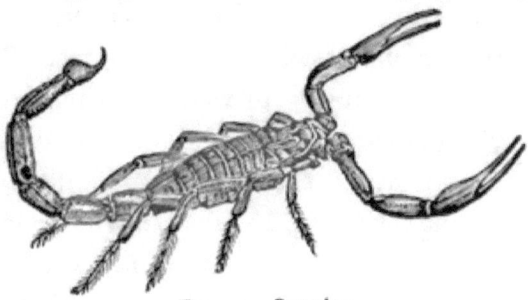

FIG. 127.—Scorpion.

joint of the abdomen bears in scorpions a sharp spine at its end, perforated by the duct of a poison-gland, whereby it inflicts painful wounds. The *Chelifer*, or Book-scorpion, sometimes found in old books, has no sting. The *Phalangers*, or Harvest-spiders, with long hooked palpi and long ungainly legs, belong to this order.

3.) *Araneida*, or Spiders, have the cephalo-thorax joined to the sac-like abdomen by a narrow constriction, and are provided at the posterior end with two or three pairs of appendages called "spinnerets." The use of the spinnerets is to reel out the silk for their web from the silk-glands. The tip of each is perforated by many pores, through which the silk escapes, so that each thread of the web may consist of several hundred strands. The silk is fluid at first, but rapidly hardens. The hind feet have comb-like claws for pressing the silk together. Sometimes one pair of the hinder appendages consists of palpiform organs. The mandibles are vertical, and end in a powerful hook. The maxillæ, or palpi, which in Scorpions are powerful claws, in Spiders resemble tho-racic feet. The brain is of large size, and the nervous system greatly concentrated.

The instincts of Spiders are very remarkable. They are the most wily of Articulates. They display great skill and industry in weaving their webs, and some species (called Mason-spiders) excavate cavities in the ground, which they line with a silken web, and close the entrance with a lid which moves upon a hinge.

5. MYRIAPODA (*myrios*, numerous; *pous*, foot) is a small class, including the Centipedes and the Millipedes. The body is divided into segments, twenty or more, to each of which legs are appended. They resemble Worms in their form, and in the simplicity of their nervous and circulatory systems; but the skin is hardened by chitine and the legs are articulate. They breathe by trachea, or tubes, have two antennæ, and a variable number of eyes.

1.) *Chilognatha*, (*cheilos*, lip; *gnathos*, jaw.) This order contains the Thousand-legged Worm, (*Julus.*) The body is round, legs very numerous, sometimes a hundred pairs, each segment having two pairs. Mouth without palpi. Lower lip composed of confluent maxillæ. They are of slow locomotion, harmless, and vegetarian.

2.) *Chilopoda*, (*cheilos*, lip; *pous*, foot,) are characterized by a flat body, with fifteen to twenty pairs of legs. The mouth possesses a hollow duct for the passage of fluid from a poison-gland. The terminal section of the body

carries a pair of spines.   Sometimes the tail is curved
into a formidable poisonous hook, as in the Scorpion.
In temperate climates the Chilopoda are harmless, but
often dangerous in hot countries.   Such is the Centipede,
(*Scolopendra.*)

6. INSECTA.   This class contains more species than
all the rest of the Animal Kingdom, 150,000 having
been already described.   Its typical character is well

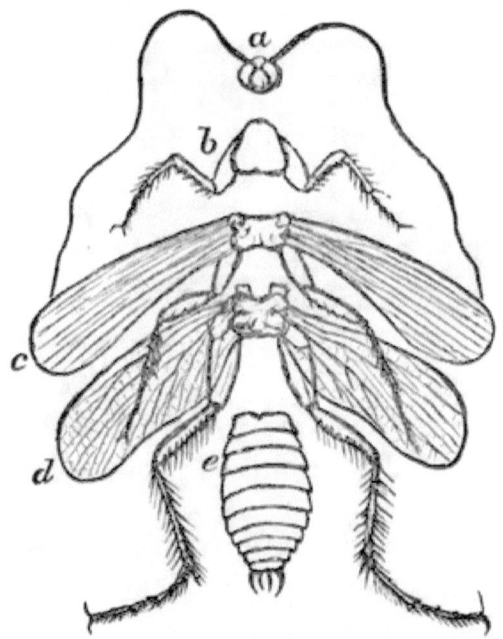

FIG. 128.—Diagram of Insect, (*Blatta orientalis.*) *a.* Head with compound eyes and
antennæ. *b.* Prothorax with first pair of legs. *c.* Mesothorax with second pair of legs
and first pair of wings. *d.* Metathorax with third pair of legs and second pair of wings.
e. Abdomen without limbs, but carrying terminal appendages, which are subservient in
reproduction.

marked, yet it contains a large number of instances of
special structure, arranged for evident purpose. Chap.
III, Sec. 9, and Chap. V, Sec. 6.

The body is divided into three principal segments—

a head, a threefold thorax, and a ringed abdomen.
(Fig. 128.)

The head contains the organs of sense, and supports two antennæ, which are supposed to be organs of
touch and of hearing, as well as of communication between one insect and another. There are many forms
of antennæ, but all are tubular and jointed. The eyes
are usually compound, although there are also sometimes a cluster of simple eyes, or *ocelli*. The compound
eyes have a trans-
parent surface, or
cornea, divided
into many facets,
each of the nerve-
rods having its
own pigment mass
and its own cornea.
(Fig. 129.) The

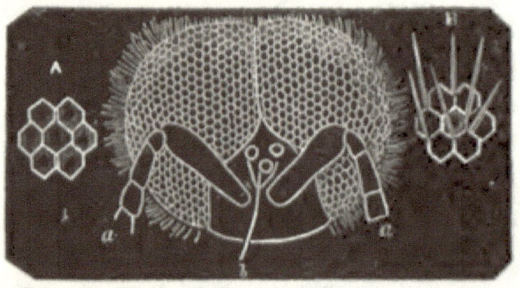

Fig. 129.—Head and Eyes of the Bee. *a. a.* Antennæ.
b. Ocelli. A. Facets enlarged. B. The same with hairs
growing between them.

common house-fly has two thousand such facets in each
eye, and in the dragon-fly there are twenty-eight thousand.

The thorax consists of three pieces, the *prothorax*,
mesothorax, and *metathorax*, each having a pair of legs;
the wings, when present, arise from the last two segments.

The abdomen contains the viscera and the organs of
reproduction. Legs are never attached to it.

The "brain," as it is called, is a mass of ganglia lying
across the upper side of the throat behind the mouth,
and the principal nerve cord, with a ganglion for each
segment, runs along the ventral side of the body.

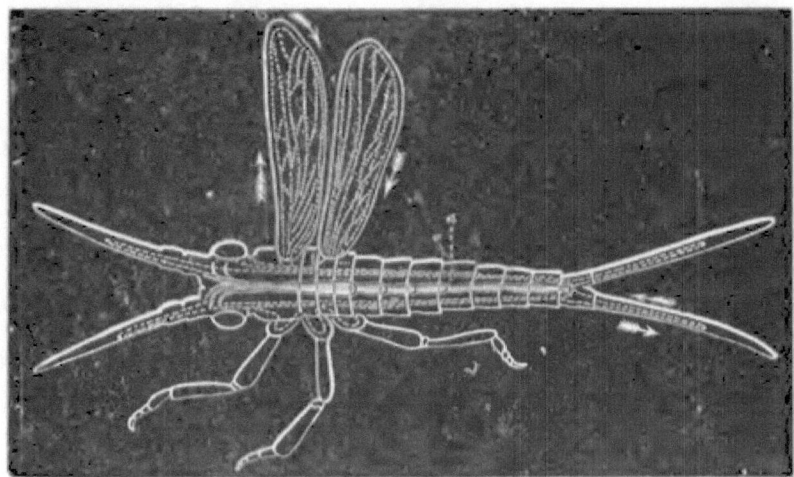

FIG. 130.—Circulation in Insects. The arrows indicate the course of the blood.

The digestive apparatus consists of pharynx, gullet, (sometimes a crop,) gizzard, stomach, and intestine. The liver is represented by tubes opening into the intestine. Many insects have glandular tubes, called

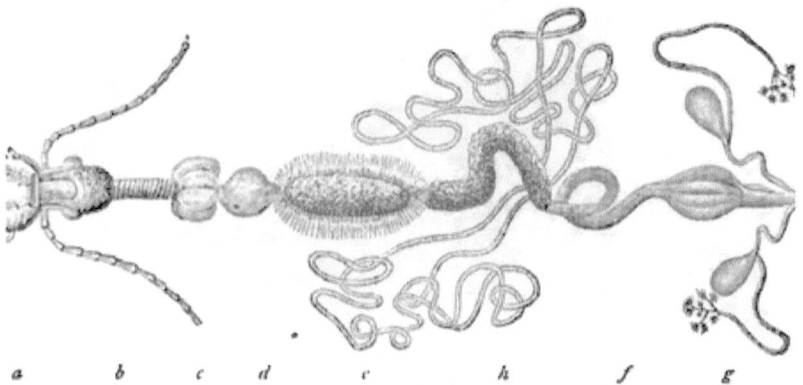

FIG. 131.—Digestive Apparatus of Beetle. *a*. Pharynx. *b*. Œsophagus. *c*. Crop. *d*. Gizzard. *e*. Chylific stomach. *f*. Small intestine. *g*. Rectum. *h*. Biliary vessels.

from their first describer, Malpighian, which open at the end of the intestine. (Fig. 131.) Some have also salivary glandular tubes and silk organs. Insects have no absorbent vessels, the chyme transuding through the walls of the canal. The blood, usually colorless, is propelled by a pulsating tube divided into valvular sacs, which allow the current to flow only toward the head. (Fig. 130.)

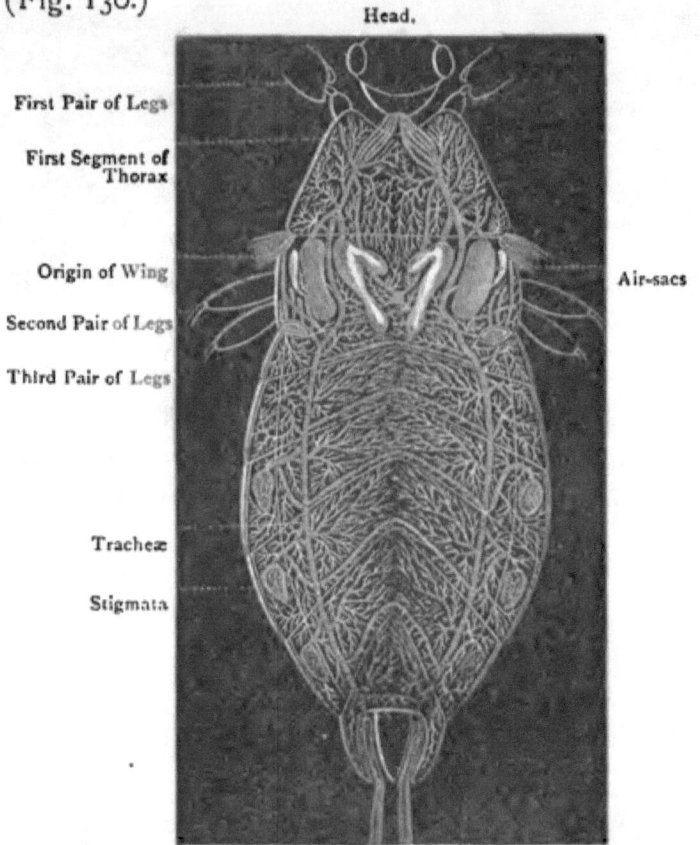

FIG. 132.—Respiratory Apparatus of Insect, (*Nepa.*)

From this tube it escapes into the cavities of the body. Respiration is provided for by a tracheal system; the air circulating in vessels instead of the blood, as in other classes. A row of apertures (*spiracles*) on each side of the body, often provided with a net-work of fibers to keep out foreign substances, communicate with branching tubes, whose membraneous wall is strengthened and kept open by a coiled spinal filament. (Fig. 132.) What are called the "nerves" of an Insect's wing are double tubes, the inner one being a tracheal branch supplying air, and the outer one, sheathing it, is a blood-vessel.

The mouth of an Insect is a very complicate apparatus. Some are *Masticatory*, or fitted for biting, as in Beetles. (Fig. 133.) Others are *Suctorial*, or for sucking, as in Butterflies. These last form a long double tube, or spiral trunk, (*proboscis*,) serving to pump up the juices of flowers. The masticating mouths consist of two pairs of horny jaws, (*mandibles* and *maxillæ*,) which work horizontally between an upper (*labrum*) lip and an under (*labium*) lip. The

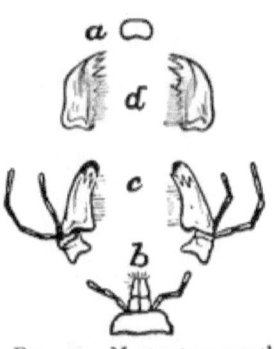

FIG. 133.—Masticatory mouth of Insect. *a*. Labrum, or upper lip. *b*. Labium, or lower lip, with jointed palpi. *c*. Maxillæ, with jointed palpi. *d*. Mandibles.

maxillæ and under lip carry sensitive jointed feelers, (*palpi*.) The front edge of the labium is generally known as the tongue, (*ligula*.)

In the Bee tribe, instead of maxillæ, we find a sheath inclosing a long, slender, hairy tongue. Entomologists have retained the same names to the different parts, under the influence of the theory of transmutation. (Fig. 134.)

The proboscis of the Fly is an enlarged lower lip, (Fig. 135;) that of the Bugs is formed by four bristles, fitted both for piercing and sucking.

Most Insects undergo metamorphosis, and exhibit four states of existence: egg, larva, pupa, and imago.

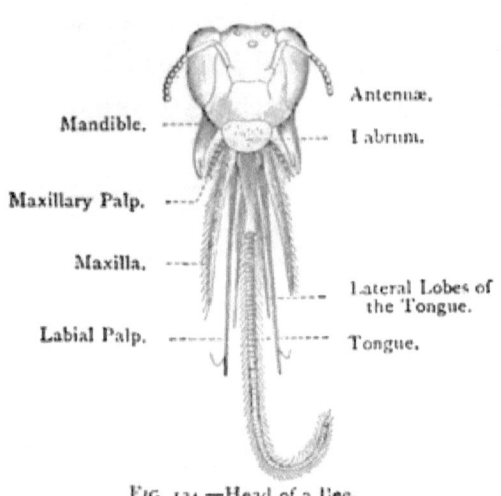

Antennæ.

Mandible.

Labrum.

Maxillary Palp.

Maxilla.

Lateral Lobes of the Tongue.

Labial Palp.

Tongue.

FIG. 134.—Head of a Bee.

The larva has little resemblance to its parent, eating and growing rapidly. It wraps itself in a cocoon and enters the pupa state, remaining apparently dead till new organs are developed, when it emerges a perfect winged Insect, or imago.

Insects have six legs, each having five parts; the *coxa* or hip, the *trochanter*, the *tibia* or shank, and the

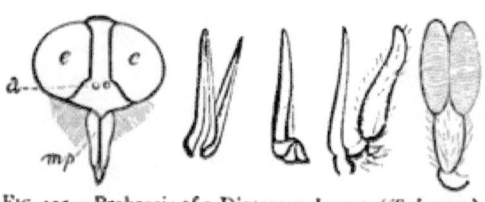

FIG. 135.—Proboscis of a Dipterous Insect, (*Tabanus.*)

tarsus. The last is subdivided into joints, generally five, and a pair of claws. Such as can walk on glass, or upside down, as the Fly, have two or three disks (*pulvilli*) between the claws. It used to be supposed that these disks acted as suckers, but it is now believed that each hair is a minute tube containing a viscid fluid by which the Fly adheres.

The male of the Great Water-beetle (*Dytiscus marginalis*) has a peculiar apparatus of suckers, large and small, on his front legs, which may be useful, but, judging from their beautiful fringes as seen under the microscope, appear rather ornamental. (Fig. 137.)

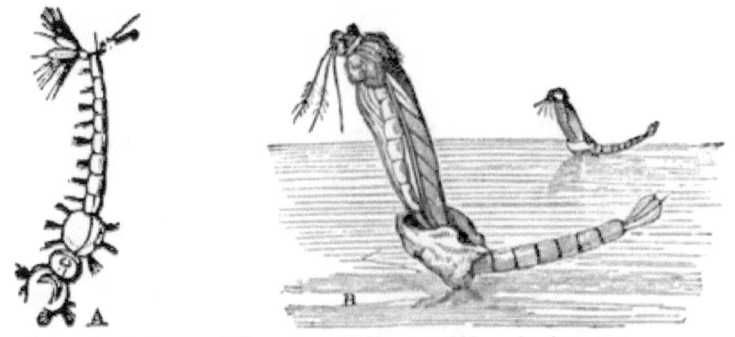

FIG. 136.—A. Larva of Mosquito. B. Escape of Mosquito from its Pupa-case

Order 1. *Neuroptera;* (*neuron*, nerve; *pteron*, wing,) includes Dragon flies, (*Libellulidæ*, (Fig. 138;) Caddis

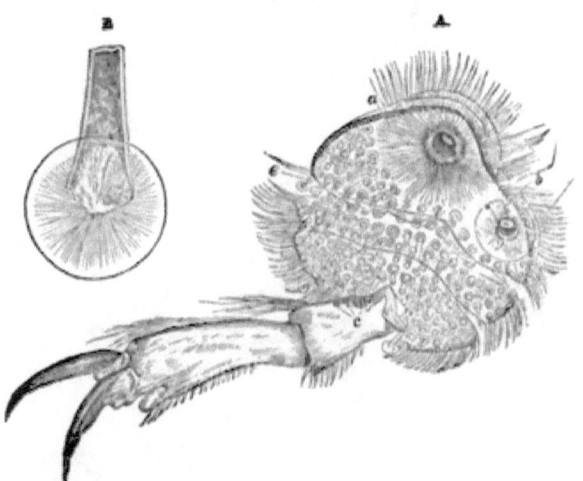

FIG. 137.—A. Foot of *Dytiscus*, showing its apparatus of suckers. *a. b.* Large suckers. *c.* Ordinary suckers. B. One of the ordinary suckers more highly magnified.

flies, (*Plioganeidæ*,) May flies, (*Ephemeridæ*,) the Ant-lion, (*Mygonalis*,) and the Termite Ants. The mouth is masticatory; wings four, equal in size, membraneous and lace-like.

20*

Order 2. *Orthoptera ; (orthos*, straight ; *ptcron*, wing,) embraces the Crickets, (*Achetina*,) Grasshoppers, (*Gryllina*,) (Fig. 138,) Locusts, (*Locustina*,) and Cockroaches,

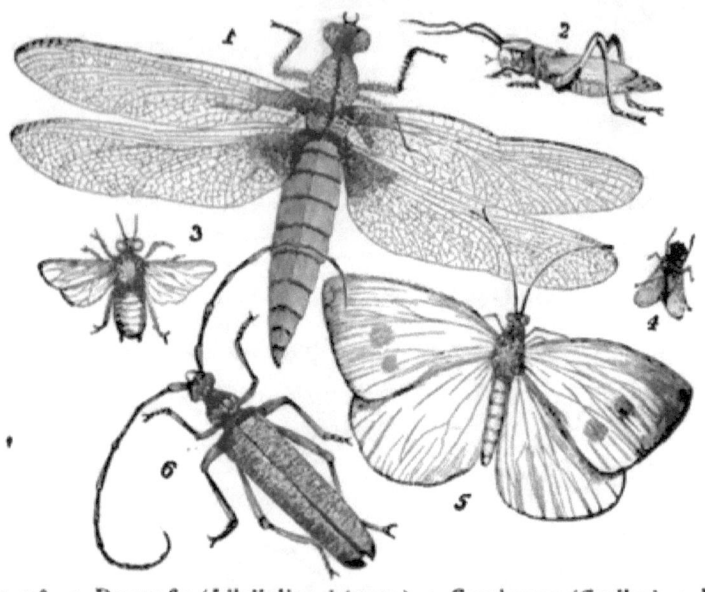

FIG. 138.—1. Dragon-fly, (*Libellulina depressa.*) 2. Grasshopper, (*Gryllus.*) 3. Bee, (*Apis mellifica.*) 4. Fly, (*Musca domestica.*) 5. Butterfly, (*Pontia brassicæ.*) 6. Musk-beetle, (*Cerambyx moschatus.*)

(*Blattina*.) Mouth masticatory. Wings four, or wanting ; anterior pair small, thickened, and overlapping along the back ; the hinder pair broad, net-veined, and folding like a fan. Legs various, being powerful jumping organs in grasshoppers, *raptorial (raptor*, a plunderer) in Mantis ; *cursorial (curro*, to run) in Locusts. Each family produces sounds which are characteristic, and which are supposed to be produced by the rapid friction of the long hind legs against the wing. The sound of the Grasshopper is said to be the highest known musical note.

Order 3. *Hemiptera*, (*hemi*, half; *pteron*, wing,) have a suctorial mouth, produced into a long, hard beak. The four wings are irregular and sparsely veined, or wanting. The body is flat above, and the legs slender. In some the four wings are opaque at the base, and transparent at the apex, whence the name of the order. Some feed on the juices of animals, and some on plants. Plant-lice, (*Aphides*,) the wingless Bed-bug, (*Cimex*,) and Louse, (*Pediculus*,) the Squash-bug, (*Coreus*,) Water-boatman, (*Notonecta*,) Seventeen-year Locust, (*Cicada*,) and the Cochineal, (*Coccus*,) are examples.

Order 4. *Coleoptera*. (*Koleos*, a sheath; *pteron*, wing.) This is the largest of the orders, containing about ninety thousand species. The thickened, horny fore-wings, or *elytra*, (*elytron*, a sheath,) are not used for flight, but to cover the hind pair. When at rest the elytra are united by a straight edge along their length, and the hind wings are folded transversely. The mouth is armed with formidable mandibles; the integument is generally hard, and the legs are strong. The larvæ are worm-like, and the pupa is motionless. The highest tribes are carnivorous. Among them we find the Tiger-beetles, (*Cicindela*,) the common Ground-beetles, (*Carabus*,) whose hind wings are often absent, the Diving-beetles, (*Dytiscus*,) with boat-like body and oar-like hind legs, the Carrion-beetles, (*Silpha*,) with black, flat bodies and club-shaped antennæ, the Goliath-beetles, (*Scarabæus*,) the Snapping-bugs, (*Elata*,) the Lightning-bugs, (*Pyrophorus*,) the spotted Lady-birds, (*Coccinella*,) the Long-horned beetles, (*Cerambycidæ*,) and the destructive Weevils, (*Curculionidæ*,) with pointed snouts. (Fig. 138.)

Order 5. *Diptera*, or two-winged Flies; have the hinder pair of wings replaced by "poisers," or "*halteres*." A few species are wingless. The eyes are large, with many facets; the tongue terminates in a fleshy knob, and the rest of the mouth is suctorial, and furnished with fine lancets; the thorax is globular, and the legs slender. The larvæ are footless grubs. Among them are House-flies, (*Muscæ*,) (Fig. 138,) Gnats, (*Culicidæ*,) Crane-flies, (*Tipulidæ*,) Forest-flies, (*Hippoboscæ*,) and Gad-flies, (*Gabrinidæ*.) The wingless Flea (*Pulex*) is also placed in this order.

Order 6. *Lepidoptera*, (*lepis*, scale; *pteron*, wing,) includes Butterflies and Moths. They have four large wings, thickly covered on both sides with minute over-lapping scales, of different colors, and often arranged in patterns of exquisite beauty. These scales are epidermic appendages of a similar nature to hairs, and every family has a special form of scale. The head is small, the body is cylindrical, and the legs are little fitted for locomotion. The mouth is a proboscis, or coiled tube, sometimes an inch long. The caterpillar, or larva, has a worm-like form, and from one to five pairs of abdominal legs, in addition to the six on the thorax. The mouth is formed for mastication.

There are three groups: the gay Butterflies, (Fig. 138,) having knobbed or hooked antennæ, flying in the sunshine only, and keeping their wings vertical when at rest; the dull-colored Sphynges, or evening Moths, with antennæ thickened at the middle, and flying at twilight; and the nocturnal Moths, whose antennæ are thread-like and often feathery, and which prefer the night.

The pupa of Butterflies is unprotected, and is generally suspended by a silken thread. The pupa-case is generally ornamented with golden spots, hence the common name, *chrysalis*. The pupa of Moths is inclosed in cocoons.

Order 7. *Hymenoptera*, (*hymen*, membrane; *pteron*, wing,) includes Bees, (Fig. 138,) Wasps, Ichneumons, Saw-flies, and Ants. The mouth is fitted for both biting and suction; the legs are for locomotion as well as support; and the four membraneous wings are equally transparent, and interlock by small hooks during flight. The females are usually provided with a sting, or borer. The larvæ are footless, helpless grubs, generally nurtured in cells, or nests.

The colony of Bees is formed of the perfect female, called the "Queen-bee," many perfect males, or drones, and a swarm of sexless bees, the neuters, or workers. The drones and the neuters are produced by parthenogenesis. (Chap. III, Sec. 16.)

The "vespiary" of the Wasps, like the hive of the Honey-bee, contains males, females, and neuters; but the perfect males work equally with the neuters.

Ants (*Formicidæ*) also form colonies, and the observations made upon many species show a wonderful amount of intelligence in these creatures. In many ant-colonies the neuters consist of two classes—"the workers," who do all the building and storing of the little town, and "the soldiers," who defend the works. Their treatment of Plant-lice, or Aphides—keeping them, and milking them as men do cows—their slave-capturing expeditions, and the recently-discovered agricultural ant-colonies,

furnish abundant food for the propensity to the marvelous in human nature, and prove to the philosophic
observer of creation how closely related are all living
things to properties of thought, affection, and will,
which are generally regarded as spiritual ratl.er than
material.

CHAPTER XV.

VERTEBRATA.

Thus the seer,
With vision clear,
Sees forms appear and disappear,
In the perpetual round of strange
Mysterious change
From birth to death, from death to birth,
From earth to heaven, from heaven to earth;
Till glimpses more sublime
Of things unseen before,
Unto his wondering eyes reveal
The Universe, as an immeasurable wheel
Turning for evermore
In the rapid and rushing river of Time.
—LONGFELLOW.

1. THE type, or sub-kingdom, Vertebrata, (*vertebra*, a joint of the back, from *vertere*, to turn,) is characterized by the separation of the greater part of the nervous system from the general cavity of the body. A transverse section of the body exhibits two cavities, or tubes; the dorsal, or neural, tube, containing the cerebro-spinal nervous system, and the ventral, or hæmal tube, inclosing the alimentary canal, heart, lungs, and a double chain of ganglia belonging to the sympathetic system of nerves. The ventral cavity, with its contents, corresponds to the whole body of an Invertebrate animal, while the dorsal tube is distinctly typical.*

Vertebrates have an internal, jointed skeleton, capable

* See Frontispiece.

of growth and repair. (Chap. IV, Sec. 13, *d*.) During embryonic life it is represented by the *notochord*, a fibro-cellular rod, tapering to either end, but this is replaced by a more highly developed column of cartilage or bone, except in the doubtful *Amphioxus*. The column and cranium are never absent, although other parts may be wanting, as the ribs in Frogs, limbs in Snakes, etc. The limbs never exceed four, and when present, are always articulated to the internal skeleton, on the ventral side of the body, while the limbs of Invertebrates are developed from an external skeleton, on the neural side. The

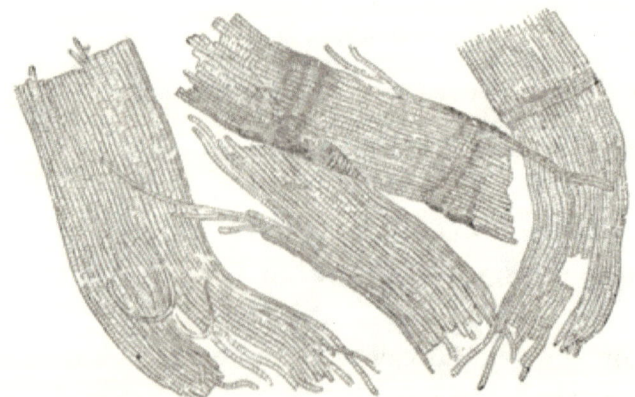

FIG. 139.—Muscular Fibers.　Magnified 200 diameters.

muscles moving the limbs are attached to the endoskeleton and not to the exoskeleton, as in Invertebrates. Muscular tissue is found in all animals, from Radiates to Man. The most complete development of muscles is in the *Pentacrinus*. (Chap. XII, Sec. 5.) Voluntary muscular tissue always has a transversely striated appearance under the microscope, (Fig. 139,) while those fibers not under the control of the will are smooth.

The circulation of the blood is complete in Vertebrates, the arteries being joined to the veins by capillaries, so that the blood never escapes into the visceral cavity, as in the Invertebrates. All have a portal vein, carrying blood through the liver, and all have lacteals and lymphatics. The blood is red, and contains both red and white corpuscles. The teeth are developed from the dermis, never from the cuticle, as in Mollusks and Articulates; the jaws move vertically, and are never modified limbs. The liver and kidneys are always present. The respiratory organs are either gills or lungs, or both. Vertebrates are the only animals which breathe through the mouth.

The arrangement of the circulatory system varies in the different classes, in accordance with the structure of the respiratory organs. In Fishes (Fig. 140) the heart is double as in the Oyster, but instead of driving arterial blood over the body, it receives the returning, or venous blood, and sends it to the gills. From the capillaries of the gills the blood is passed into a large artery, or *aorta*, along the back, which distributes it by a complex net-work of capillaries among the tissues. These capillaries unite with the ends of the veins which pass the blood into the auricle of the heart, after purification in the liver and kidneys.

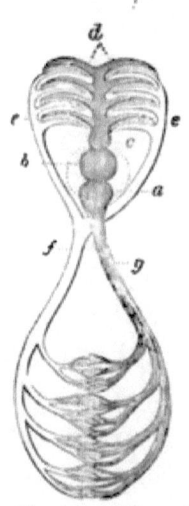

Fig. 140. — Plan of Circulation in Fishes: *a*. Auricle. *b*. Ventricle. *c*. Pulmonary Artery. *e*. Pulmonary Veins, bringing blood from the gills, *d*, and uniting in the Aorta, *f*. *g*. Vena Cava.

In Amphibia and Reptiles (Fig. 141) the heart has three cavities; two auricles

21

and one ventricle. The venous blood from the body is received into the right auricle and the purified blood from the lungs into the left. Both communicate with the ventricle, which pumps the mixed blood part to the lungs and part around the system.

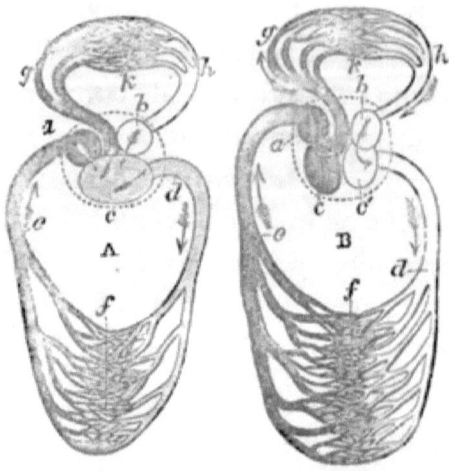

Fig. 141.—A. Plan of Circulation in Amphibia and Reptiles. B. Plan of Circulation in Birds and Mammals. *a.* Right Auricle receiving venous blood from the system. *b.* Left Auricle receiving arterial blood from the lungs. *c. c'.* Ventricles. *d. e. f.* Systemic Artery, Vein, and Capillaries. *g. h. k.* Pulmonary Artery, Vein, and Capillaries.

The highest form of circulation is seen in the warm-blooded Vertebrates, Birds, and Mammals. The heart has four cavities — a right and left auricle, and a right and left ventricle. The right auricle receives the blood from the veins, transmits it to the right ventricle, which sends it to the lungs. The left auricle receives it from the lungs, and sends it to the left ventricle, which propels it over the body. The two auricles contract together, and so also do the ventricles, making certain faint sounds, which may be imitated by the words *lubb tup.* (Fig. 141.)

The greatest differences between Vertebrates and other animals are found in the Nervous system, which, as we have seen, has a distinct tube or cavity in this type, altogether unlike the plan of structure elsewhere.

In living things, like the Protozoa, or Protophytes, which are composed of a simple mass of bioplasm, all the functions necessary for animal or vegetable life belong equally to every atom of the mass. In Chap. V, Sec. 7, it was shown that the simplest plants and animals differ from the highest, or more complex, only in the modification of some parts of the structure to serve special functions. Thus locomotion is served by the change of bioplasm into muscle, or bone, external protection by transformed epidermal bioplasm, as described in Chap. IV. To regulate and harmonize the complex organs of digestion, respiration, circulation, and secretion, and to conduct sensation and motor force, seems to have been the object of the change of bioplasm into nervous matter.

Nerve matter exists in the form of cells and of fibers. The cells are soft and grayish, and are generally found accumulated in masses or *ganglia*, sometimes called nerve-centers. The fibers are of two kinds, one soft and nucleated, the ganglionic or sympathetic fibers, and ordinary nerve-fibers.

These latter are for a great part of their length inclosed in a transparent sheath, which coagulates after death into a white substance—the white substance of Schwann. A number of these fibers, thus ensheathed, are bound in bundles, which are called *nerves*. Some of these fibers proceed or conduct impressions from the surface, or from the different organs where they are found, toward the gray centers only, and are called *afferent* or sensory nerves. Others conduct an influence from the centers to contract or move the muscles, and are called *efferent*, or motor nerves. Thus, on receiving any impression, as the prick of a pin, an afferent nerve conducts the impression to the center, from which an

efferent nerve conducts power for the muscles to contract.
If the afferent nerve of a part is cut across or injured,
sensation is lost, but motion remains; but if the efferent
nerve is cut, the power of motion is lost while sensibility
continues. This form of nerve-action is called *reflex*.
Many actions of this sort are wholly involuntary, as the
motions of the limbs in paralytics excited by tickling the
soles of the feet.

In the Star-fish we find a nervous ring around the
mouth, made of five ganglia, with radiating nerves, corresponding with the type of structure. The Mollusks
have an irregularly scattered nervous system, consisting
of two or more ganglia around the gullet and one or two
more in the posterior region, united by threads, and
sending fibers to various organs. The Articulates have
generally a double nervous cord along the ventral side,
studded with ganglia of nearly uniform size, except the
first, which is largest. In the higher forms, as the Bee,
several ganglia are fused together in the head and thorax, indicating a concentration of organs for sensation
and locomotion.

The nervous system of the Invertebrates is *homolog-
ically* represented by the ganglionic or sympathetic system of Vertebrates, which supplies the unstriped or involuntary muscles, and presides over organic or visceral
functions, such as digestion and circulation. In addition
to the sympathetic system, Vertebrates have a brain and
spinal cord, forming the cerebro-spinal system, (Fig. 142,)
to which there is nothing similar in other animals, and
which presides over what are called the functions of animal or sentient life, as sensation and locomotion. Yet

as many Invertebrates exhibit sensibility and voluntary actions, it follows that *analogically* the nervous system in them represents both the sympathetic and cerebro-spinal systems of Vertebrates.

The form of the brain differs much among Vertebrates. In some the cerebral hemispheres are small; in certain Fishes smaller than the optic lobes; in the higher tribes they nearly or quite overlap both olfactories and cerebellum. The brain may be smooth, as in most cold-blooded animals, or greatly convoluted, as in Man.

Vertebrates are subdivided into five classes: *Fishes, Amphibians, Reptiles, Birds,* and *Mammals.* The first three are cold-blooded, the other two warmblooded. Fishes and Amphibians agree in having gills during all or a part of their lives. The rest never have gills.

2. FISHES, (*Pisces.*) These are considered the lowest of Vertebrates, yet they are so numerous as to embrace nearly one half of all Vertebrates, and so varied that it is difficult to frame a definition which shall include them all.

Fishes live in the water and breathe

21*

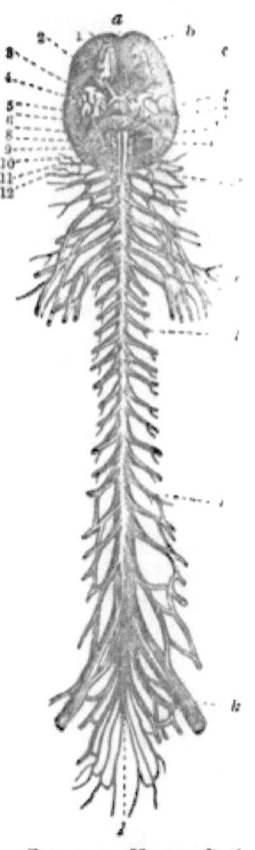

FIG. 142.—Human Brain and Spinal Cord, one fifth natural size. *a.* Great longitudinal fissure. *b.* Anterior lobe. *c.* Middle lobe. *d.* Medulla oblongata. *e.* Cerebellum. *f.* First spinal nerve. *g.* Brachial plexus of nerves supplying the arms. *h.* Dorsal nerves. *i.* Lumbar nerves *k.* Sacral plexus of nerves for the limbs. *l.* Cauda equina. The figures indicate the twelve pairs of cranial nerves, of which one is olfactory, two are optic, and eight auditory.

by means of gills. They are generally covered with scales, and they have fins instead of limbs. They have large immovable eyes, but no external ears. Both jaws are movable. The teeth are numerous, and are generally recurved spines, as in the Pike; flat and triangular, with serrated edges, in the Shark; and tessellated, in the Ray. The digestive tract is relatively shorter than in other Vertebrates. The blood is red, and the heart has two cavities, an auricle and a ventricle, both on the venous side. Ordinary fishes have four gills, the water escaping by one external aperture, or "gill-slit;" but in Sharks there is a separate opening for each gill.

There are four principal varieties of fish-scales. (Fig. 143.) 1. Cycloid scales, (*cyclos*, a circle,) which are most common; thin, flexible, and silvery, as in the Salmon. 2. Ctenoid, (*kteis*, a comb,) with a comb-like fringe of toothed processes. 3. Ganoid, (*ganos*, brightness,) generally larger than

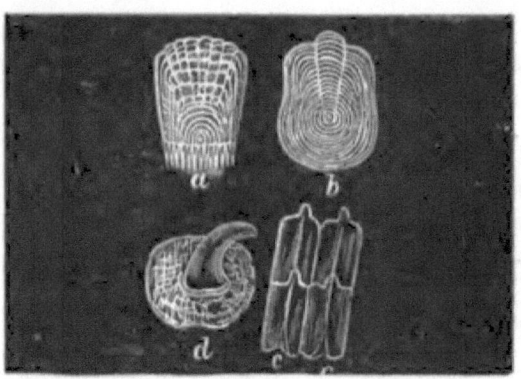

FIG. 143.—Varieties of Fish Scales. *a*. Ctenoid scale. *b*. Cycloid scale. *c*. Ganoid scale. *d*. Placoid scale.

the preceding, and having an under layer of bone with a superficial layer of enamel. Most ganoid fishes are extinct. 4. Placoid, (*plax*, a flat plate;) these are formed of bony granules, or tubercles, or plates, the plates often being furnished with spines.

Most fishes have a series of small scales running along

the side of the body, called the lateral line. Each scale is perforated by a tube which runs along the whole length of the body, and is connected with cavities in the head which secrete the mucus for lubricating the scales, and enabling the fish to move with little resistance.

Order 1. *Pharyngobranchs*, (*pharynx*, the pharynx, and *bragchia*, gills.) This contains but one member, the Lancelet, (*Amphioxus lanceolatus*,) which burrows in the mud of the Mediterranean Sea. It is such an eccentric creature, without skeleton, limbs, brain, heart, lymphatics, or red blood, that it can hardly be considered a Vertebrate at all. Yet, because it has a persistent notochord, evolutionists have made much ado over it, and it figures largely in their imaginary *Phylogenies*. (Chap. III, Sec. 8.)

Order 2. *Marsipobranchs*, (*marsipos*, a pouch.) They have a cartilaginous skeleton and sac-like gills, but no scales, limbs, or lower jaw, and only one nasal organ. They comprise the eel-like Lampreys and Hags. (Fig. 144.) The mouth is round and sucker-like; and in the

FIG. 144.—Lamprey.

Hags (*Myxinæ*) contains a single large recurved, serrated fang for piercing the bodies of their prey. Respiration is carried on by muscular little pouches (*marsupia*) placed on the sides of the neck.

Order 3. *Teleosts*, (*teleios*, perfect ; *osteon*, a bone,) includes all the true osseous fishes. The skull is complicated, the upper and lower jaws complete, and the gills are comb-like, or tufted. The tail is *homocercal*, having equal lobes. The other fins vary in number and position.

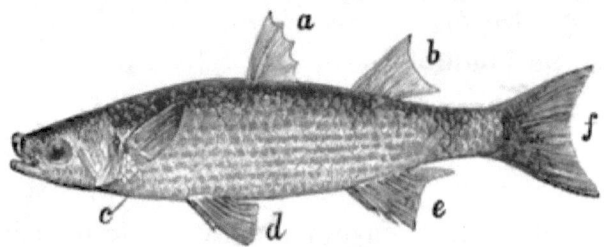

FIG. 145.—Gray Mullet: *a*. First dorsal fin. *b*. Second dorsal. *c*. Pectoral. *d*.Ventral. *e*. Anal. *f*. Caudal.

In the soft-finned Fishes, the ventral fins (Fig. 145) are absent, as in the Eels ; or attached to the abdomen, as

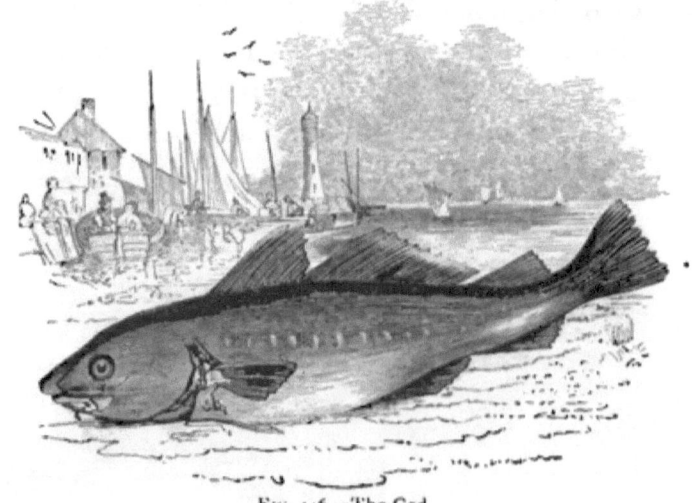

FIG. 146.—The Cod.

in Salmon, Herring, Pike, and Carp ; or placed under the throat, as in the Cod, (Fig. 146,) Haddock, and Flounder.

In the spring-finned Fishes, the ventrals are generally under or in front of the pectorals, and the scales are ctenoid, as in the Perch, Mullet, and Mackerel.

Order 4. *Ganoids* include the Sturgeons, (Fig. 147,)

FIG. 147.—The Sturgeon, (*Acipenser Sturio.*)

Bony-pike, Polypterus, and many extinct forms. The skeleton is rarely completely ossified; the ventral fins are placed far back, and the tail is *heterocercal*, or unequally lobed, from the vertebra continuing in the upper lobe.

Order 5. *Elasmobranchs* (*elasma*, a thin plate) contain Sharks, Rays, and Chimeræ. The gills are formed of thin laminæ, arranged like the leaves of a book. They have a cartilaginous skeleton, and a harsh skin called "shagreen." The gill-openings are uncovered, and the mouth is generally under the head, (except in the Chimeræ.) The ventral fins are placed far back, the pectorals are large, and in the Rays enormously developed and the tail is heterocercal.

Order 6. *Dipnoi*, (*dis*, twice ; *pnœ*, breath,) comprises the Mud-fishes (*Lepidosiren*) of tropical rivers. (Fig. 148.)

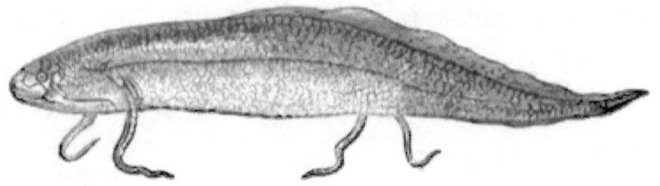

FIG. 148.—Lepidosiren.

They have eel-like bodies covered with cycloid scales. Both ventral and pectoral fins are present, but are small filiform organs, nowise resembling ordinary fins. They have rudimentary external gills, and internal ones which communicate with the exterior by a single slit. They also possess true lungs, which communicate with the gullet by a tube or trachea. The heart has two auricles and one ventricle. They are quite Amphibian in structure, and live long out of the water.

3. CLASS II, AMPHIBIA, (*amphi*, both ; *bios*, life,) receives its name from the animals it contains being able to live both on land and water. They are cold-blooded Vertebrates which breathe by gills during some part of their life, but sooner or later possess lungs. Some retain their gills through the whole of their life, as the Proteus, Siren, and Axolotl ; others lose their gills after a time, and breathe by lungs only, as Frogs, Toads, and Newts. All undergo metamorphoses after leaving the egg, passing through the "tadpole" state, in which they resemble Fishes in their respiration, circulation, and locomotion.

Order 1. *Urodela*, (*oura*, a tail ; *delos*, visible,) the tailed Amphibia. They have a naked skin, and two to four legs. The aquatic Newts and land Salamanders drop their gills, and always have four limbs.

Order 2. *Labyrinthodontia* (*labyrinthos*, a labyrinth, *odous*, a tooth) are now all extinct. They resembled gigantic Salamanders, except in their complex teeth and exoskeleton of bony plates.

Order 3. *Gymnophiona*, (*gymnos*, naked ; *ophis*, a snake,) also called Cecilia. They have neither tail nor limbs, a snake-like form, minute scales in the skin, and numerous ribs.

Order 4. *Batrachia*, or *Anoura*. (Fig. 149.) (*Batrachos*, frog ; *ana*, without ; *oura*, a tail.) These are tailless

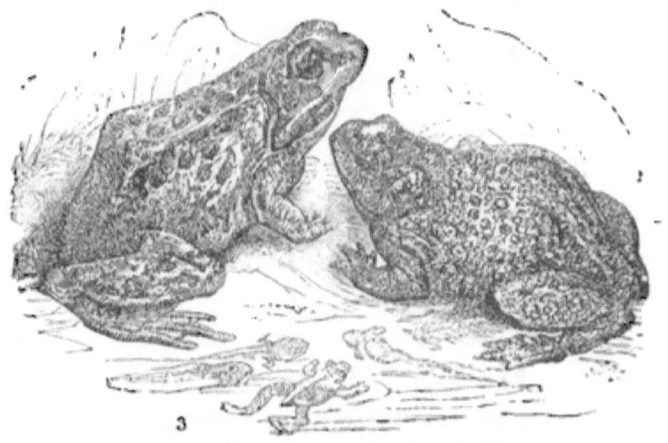

FIG. 149.—1. Frog, (*Rana temporaria*.) 2. Toad, (*Bufo vulgaris*.) 3. Tadpoles.

Amphibia, and comprise Frogs and Toads. They have a naked, moist skin, ten vertebræ, and no ribs. They have four limbs, the hinder longer than the fore-legs. They have four fingers and five toes. The tongue is long, fixed at the anterior, and doubled up. It can be thrown out rapidly as an organ of prehension. The eggs are laid in the water, enveloped in a glairy mass, and the tadpoles are like the Urodelans till the gill and tail are absorbed. Frogs (*Rana*) have teeth in the upper jaw, and webbed feet. Toads (*Bufo*) have neither teeth nor webbed feet.

4. Class III. REPTILIA, or Reptiles. These are air-breathing, cold-blooded Vertebrates, differing from Fishes and Amphibians by never having gills, and from Birds by being covered with horny scales, or bony plates. The skeleton is ossified, and never cartilaginous. Most are carnivorous, and teeth are present, except in Turtles, where a horny sheath covers the jaws. The lungs are imperfectly cellular, and the heart is three-chambered, containing two auricles and one ventricle, which is some-times divided by a partition. In all cases a mixture of arterial and venous blood is circulated. The limbs, when present, have three or more fingers as well as toes.

There are four orders of living and five of extinct Reptiles. The living orders are Snakes, Lizards, Tur-tles, and Crocodiles.

1.) *Ophidia*, or Snakes. (Fig. 150.) These have no visible limbs, but a vast number of vertebræ. The Py-thon has two hundred and ninety-one, the Rattlesnake one hundred and ninety-four, and the Boa Constrictor three hundred and five. They have immovable trans-parent eyelids. The tongue is bifid (cleft) and extensile. The mouth is very dilatable, from the number of joints in the lower jaw united only by ligament. The skin is shed in one piece by reversing it. Snakes move well either on land or in water.

Poisonous snakes, as Vipers and Rattlesnakes, usually have a triangular head covered with small scales, a constriction, or neck, behind the head, two or more fangs and few teeth, small eyes with vertical pupil, and short, thick

FIG. 150.—1. Rock Snake, (*Python molurus*.)   2. Spectacled Snake, (*Cobra de Capello*.   3. Boa Constrictor.

tail.   In the harmless Snakes the head blends with the neck, and is covered with plates, (Fig. 151,) the teeth are numerous in both jaws, the pupil is round, and the tail tapering.

2.) *Lacertilia*, comprising Chameleons, Blindworms, and Lizards, are distinguished from the other orders by possessing clavicles, and having teeth not lodged in sockets, as in the Crocodiles.   They are often like Snakes with four limbs, each having five digits.   Some have no exoskeleton, but it is generally present in the shape of

22

scales, or horny plates, or spines.   In the *Iguanidæ* it is
elevated into a crest, or mane, of horny scales, covering
also the throat-pouches.   The *Draco volans*, or Flying
Lizard, has a cutaneous expansion from the false ribs
which enable it to take short flights through the air.

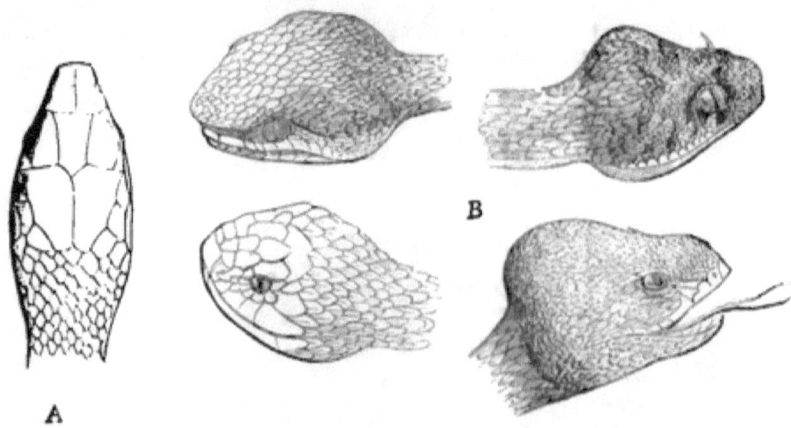

FIG. 151.—A. Head of Harmless Snake.   B. Heads of Poisonous Snakes of different
genera.

The tongue is bifid in many of this order, but in Chameleons it is a long, round, muscular organ, clubbed at
the end, and coated with a viscid secretion, by means of
which it catches great numbers of flies by shooting it
out with remarkable speed.

3. *Chelonia*, or Tortoises and Turtles.   These resemble
Amphibians in some respects, but their structure is very
peculiar.   The exoskeleton unites with the endoskeleton,
forming the *carapace*, or case, which includes the viscera
and muscular system.   The vertebræ are soldered together and the ribs are expanded, making the walls of
the carapace.   The ventral piece is called the *plastrom*,
or sternum.   (Fig. 152.)

The Sea-turtles, as the edible Green Turtle and the Hawk's-bill Turtle, which furnishes the tortoise-shell of commerce, have the limbs formed for paddles. The fresh-water forms, as the Snapping Turtle, (*Chelydra,*)

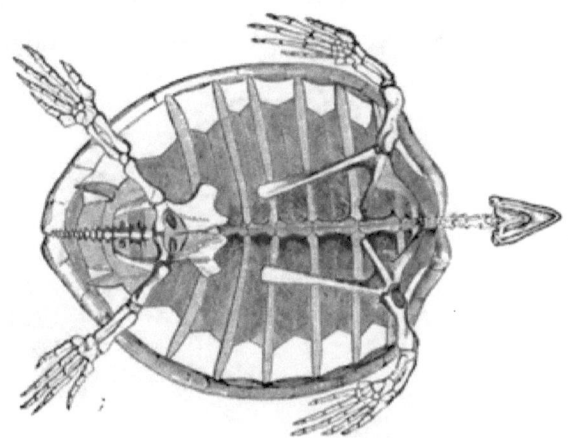

FIG 152.—Skeleton of Turtle.

are amphibious, and have palmated feet. Land Tortoises (*Testudo*) have short, clumsy limbs, fitted for slow motion on land.

4. *Crocodilia,* or Crocodiles, Alligators, and Gavials. (Fig. 153.) These are the largest Reptiles. They have a double exoskeleton, one of horny scales, (epidermic,) and another of bony plates, (dermal.) The bones of the skull are united by sutures, and the jaws are furnished with numerous teeth implanted in distinct sockets. Crocodiles can be distinguished from Alligators by the fourth tooth on the lower jaw being larger than the rest, and by its projecting on each side of the snout. The toes of Gavials and Crocodiles are fully webbed, but those of Alligators are only half-webbed. Some Crocodiles in the Nile attain to a length of twenty-five feet.

The Gavials, or Crocodiles of the Ganges, have the jaws produced to an enormous length.

5. CLASS IV. AVES, or Birds, are warm-blooded Vertebrates, clothed with feathers.

The bones of Birds are very compact and ivory-like, yet lighter than those of Reptiles or Mammals. Many

FIG. 153. — 1. Crocodile, (*Crocodilus vulgaris.*) 2. Alligator, (*Alligator lucius.*) 3. Lizards.

parts of the skeleton are fused, or anchylosed together. The lumbar vertebræ are wanting, but the neck contains from nine to twenty-four vertebræ, rendering it quite flexible. The sternum is strong, and in birds of

flight possesses a median Keel, (*carina,*) to afford an
increase of space for attachment of the great pectoral
muscles. Hence these birds are called *carinatæ*. The
skull articulates with the spine by a single condyle, and
with the lower jaw by the intervention of a separate
bone, as in Reptiles.

The beak is the bird's principal organ of prehension,
and differs in shape according to habits and food. The
pharynx is simple. The œsophagus varies in different
orders. Except in some aquatic birds, the food is received first into a temporary stomach, or *crop*, which is
largest in grain-feeders. From this the œsophagus leads
to the true digestive stomach, which secretes the gastric
juice, (*proventriculus,*) and leads to the muscular stomach or gizzard, (*ventriculus bulbosus.*)

In flesh-feeders this is thin, but in grain-feeders it is a
powerful triturating organ. The small intestine is short
in carnivorous birds and long in others. The large intestine ends in a dilated sac, the *cloaca*, which also receives the terminations of the urinary and generative
organs.

The trachea is furnished with two larynges ; one at
the upper part, as is usual in animals, and one called the
syrinx, which is the principal organ of voice, at the bi-
furcation of the trachea into the two bronchi. Every
means are employed to render the respiration rapid and
complete. The lungs are large and cellular, and the
bronchial tubes not only divide continuously in them,
but conduct air into the general cavity of the abdomen
and to the interior of many of the bones.

The feathers of birds are cutaneous growths, each
22*

formed on a vascular papilla at the bottom of a pit, or follicle. They are composed, like hair, of epithelial cells. Each feather consists of a quill, or barrel, and a vane, or beard, which is composed of barbs and barbules. The

FIG. 154.—A. Ear coverts. B. Bastard wing. C. D. E. Wing coverts. F. Primaries. G. Scapulars. H. Secondaries. L. Tail coverts.

barbules, from contiguous barbs, hook into each other like the latch of a door into its catch, so as to present an even and resisting surface to the air. Feathers receive different names on different parts of the bird's body. The feathers clothing the body are called *clothing* feathers; the great quill tail feathers, so useful in steering are the *rectrices;* those lying over the humerus and scapular are the *scapulars;* the proximal end of the ulna is covered with the *tertiaries;* the distal end of the same bone with the *secondaries;* while the bones of the hand support the *primaries*, which are largest of all. Each quill often carries a little light feather just beneath the commencement of the vane, the *accessory plume*, or *plumule.* These form the greater, lesser, and under wing coverts. (Fig. 154.)

Order 1. *Natatores*, or Swimmers. These have the body boat-shaped, and the feet more or less webbed.

One division of swimming birds is called *Brevipinnatæ*, (Short-wings,) the feathers and wings being short. It

FIG. 155.—Common Tern.

includes the Penguins, Grebes, Puffins, Guillemots, and Divers. In the Penguins the wings are too short for

flight.   The legs are placed far back, and the wings
assist the webbed feet as paddles.

The Cormorants, Pelicans, Gulls, Petrels, and Terns
(Fig. 155) form the group of *Longipinnatæ*, or Long-
wings.   The beak is hooked and pointed, the tip being
often very hard.   The Albatross, one of the largest and
most beautiful birds of flight, belongs to this group.

The *Lamellirostres*, or Flat-bills, form a third division,
including Ducks, (Fig. 156,) Geese, Swans, and Flamin-

FIG. 156.—Wild Duck, (*Anas boschas.*)   North America.

goes, whose bills are horizontally compressed, covered
with a soft cuticle supplied with twigs from the fifth
nerve, and have fringed sides, which strain the muddy
food.

Order 2. *Grallatores*, or Waders, (*grallæ*, stilts,) have
long, stilt-like legs, toes free, wings large and powerful.

The Rails, Coots, Water-hens, and Jacanas form a group called *Macrodactylæ*, because the claws are very long. They are four in number, and lobed. The beak is somewhat cuneiform, and the tail is very short.

The *Cultirostres*, with elongated forcep-like bills for fishing, include the Cranes, Herons, (Fig. 157,) Stilts,

FIG. 157.—Heron.

Ibis, and Spoonbills. The legs are very long, and not covered with feathers.

The *Longirostres*, or Long-beaks, possess long and sensitive beaks, grooved by nostrils. The legs are of moderate length. They are insectivorous. The group comprises the Snipes, Woodcocks, Sandpipers, Curlews, Ruffs, and Godwits.

The Plovers, Lapwings, Bustards, Longshanks, and Oyster-catchers form the *Pressirostres*. All possess a moderate bill with a compressed tip. Feet semi-palmate, wings long and strong.

Order 3. *Cursores*, or Runners. The wings are rudimentary and unfit for flight. The legs are hollow, strong, and long. The order includes the Ostriches, Cassowaries, and Apteryx, marked by their large size, keelless breastbone, and robust legs. The African Ostrich has two toes, the Cassowary three, and the Apteryx four. The barbs of the feathers are disconnected, forming plumes.

Order 4. *Rasores*, or *Gallinaceæ*, Scratchers, or Fowls. (Fig. 158.) These have a short arched bill, and short

Fig. 158.—Turkey, (*Meleagris Gallopavo.*)

and concave wings. There are three anterior toes and one posterior. The anterior are blunt and adapted to scratching. The gizzard is very strong. The legs are usually feathered to the heel, and sometimes to the toes. The males have usually gay plumage and some

appendage to the head. Their principal food is grain.
The order comprises the common Fowl, Turkey, Par-
tridge, Grouse, Pheasant, Ptarmigan, and Pea-fowl.

The preceding orders form a legion called *Autophagi*,
since immediately on being hatched they can run about
and look after themselves. The remaining orders form
the legion *Heterophagi*, in which the young are dependent upon their mother for nourishment for some time
after birth.

Order 5. *Columbæ*, or Pigeons. These differ from the
Scratchers in possessing powerful wings. They have
slender legs, with toes ununited, and the hind toe on a

FIG. 159.—Wood-Pigeon.

level with the rest. Pigeons, Doves, (Fig. 159,) and the
extinct Dodo are found in this order.

Pigeons exhibit in a great degree the mutability of
races or varieties; all the vast number of Pigeons,

Carriers, Tumblers, and Fantails, being descended from one common stock — the blue rock Pigeon, *Columba livia*.

Order 6. *Scansores*, or Climbers, have four toes, two directed forward and two backward. They feed on insects or fruit. They are not usually musical. The majority make nests in the hollows of old trees, but the Cuckoos lay in the nests of other birds. In climbing, the Woodpeckers are aided by their stiff tail, and the Parrots by their hooked bill. Cuckoos, Parrots, Toucans, Trogons, Woodpeckers, and Wrynecks belong to this order.

Order 7. *Passeres*, or Perchers, is the most numerous of all the orders. The two outer toes are joined by membrane. Of the two others, one is always directed backward. Females are generally smaller than males, and have more somber colors. Their nests are often of beautiful construction. The voice is often musical, the plumage lustrous, and the power of flight perfect.

The *Conirostres*, (Cone-bills,) with a short, strong, roundish or conical beak, which tapers rapidly from a broad base to a short tip, includes the Finches, with the Sparrows, Larks, Crossbills, Crows, and Hornbills. Birds of Paradise, also, and many migratory birds, as the Starling, belong here.

The Shrikes, Fly-catchers, Nightingales, Orioles, Robins, Thrushes, Tits, and Warblers form another group, the *Dentirostres*, or notched-beaks, from having an abrupt notch on the margin of the upper beak, near its tip.

The Humming-birds, Hoopoes, Wrens, Creepers, and

Honey-eaters, constitute the *Tenuirostres*, (Slender-beaks,) in which group the beak is elongated into a slender forceps for extracting honey or insects from the deep parts of flowers. The plumage is often of a gorgeous metallic luster.

The Swallows, (Fig. 160,) Martins, Goatsuckers, King-fishers, and Swifts, make up the *Fissirostres*, (Cleft-beaks,)

FIG. 160.— Swallow, (*Hirundo.*)

with a wide but short beak. During flight the mouth is kept wide open, and any insects it encounters are retained by a viscid secretion. A young Swallow will in this way consume over a thousand flies and gnats in a day.

Order 8. *Raptores*, or Raveners. These are readily recognized by their beak, which is a formidable weapon with sharp edges and an acute hooked tip. The upper bill overlaps the lower. The toes are three in front and one behind, all terminated by sharp hooked talons.

23

Wings, very large and powerful. Legs, short, stout, and strong.

Fig. 161.—Golden Eagle.

Fig. 162.—Barn Owl.

There are two sections: the *Diurnal*, whose bright eyes are on the sides of the head, wings pointed, and metatarsus and toes covered with scales, as Vultures, Kites, Hawks, Falcons, and Eagles, (Fig. 161;) and the *Nocturnal*, whose large eyes are directed forward, and surrounded with radiating feathers, metatarsus feathered, plumage soft, as Owls. (Fig. 162.)

6. CLASS V. MAMMALIA, or Mammals. These are warm-blooded Vertebrates possessed of mammary glands. They suckle their young. The thorax and abdomen are separated by a diaphragm, the red corpuscles of the blood are doubly concave and round, (except in the Camel and the Llama,) and either a part or all of the body is hairy.

All Mammals use their lips for prehension, which are assisted in some orders by their fore-limbs. The Carnivora tear their prey with their claws, but do not use them as prehensile organs. The proboscis of the Elephant, the snout of the Tapir, the long viscid tongue of the Ant-eater, and the long tongue of the Giraffe, are special prehensile organs.

The teeth of Mammalia differ in the different orders, as to number, size, and shape. The true Ant-eater has no teeth, the Narwhal has but two, one of which is rudimentary, but the Dolphin has one hundred and ninety. The Whalebone-whale (*Balæna mysticetus*) has, instead of true teeth, a series of plates of whalebone ranged in rows along the upper jaw. From these plates a long fringe of whalebone threads hangs down, which acts as a sieve in straining the water from the myriads of little mollusca which constitute the chief food of the whale.

There are three distinct types of stomach among Mammals: the simple, the compound, and the complex stomach. The simple stomach is a single cavity lined by epithelium, which secretes gastric juice. The compound stomach has the cavity divided by folds into two or more spaces. The tissue-elements, however, are the same throughout. The complex stomach is peculiar to the Ruminants. It consists of four cavities: the *paunch*

—which is the largest cavity of all—to store the food and mix it with the water it contains, and which in Camels, Llamas, and Dromedaries contains pits closed by muscular rims for storing up fluid when the animal is going a long arid journey; the *reticulum*, or honey-comb apartment, where the food is made into small round pellets, to be regurgitated into the mouth, where they undergo a second mastication; the *manyplies*, with its mucous membrane arranged in parallel folds, like the leaves of a book, and where some digestion of soluble parts of the food may occur; and the *rennet*, or true digestive stomach, where the albuminous principles of the food are extracted and absorbed by the veins.

The digestive canal is much longer in herbivorous than in carnivorous Mammals, being thirty times the length of the animal in the sheep, and five times the length in the cat and dog.

An external ear is rarely absent; the eyes are always present, though rudimentary in some burrowing animals; and, while in all other animals the embryo is developed from the nourishment contained in the egg, in Mammals it derives its support, almost from the beginning, directly from the parent, and, after birth, is sustained for a time by the milk secreted by the mammary glands.

Order 1. *Monotremata.* Includes two singular forms, the Duck-mole, (*Ornithorhynchus*,) and Spiny Ant-eater, (*Echidna*,) both confined to Australia. The former has a fur covering, a bill like a Duck, and webbed feet. The latter is covered with spines, has a long, toothless snout, like the Ant-eaters, and the feet are not webbed. Both burrow, and feed upon insects.

Order 2. *Marsupiala*, (*marsupos*, a pouch,) comprises Kangaroos, Phalangers, Wombats, and Opossums, (Fig. 163.) Except the latter, all are restricted to Australia

FIG. 163.—Virginian Opossum.

and adjacent islands. The young are always born prematurely, and are transferred by the mother to a pouch on the abdomen, where they are attached to the nipples, and the milk is forced into their mouths by special muscles.

Order 3. *Edentata*, (toothless.) This order contains very diverse forms, as the leaf-eating Sloths and the insectivorous Ant-eaters and Armadillos (Fig. 164) of South America, and the Pangolin and Orycteropus of the Old World. The gigantic fossils Megatherium and Glyptodon belong here. The Sloths and Ant-eaters are covered with coarse hair; the Armadillos and Pangolins with an armor of plates, or scales. The Ant-eaters and Pangolins are strictly edentate, or toothless; the rest have

23*

molars, sometimes very numerous, wanting, however, enamel and roots.

FIG. 164.—Armadillo.

Order 4. *Rodentia*, (gnawers.) These have two long curved incisors in each jaw, which serve for gnawing the bark of trees, or other substances on which the Rodent feeds. The front only is covered with enamel, and the rest of the tooth is composed of softer dentine, which, wearing faster than enamel, leaves a chisel-like edge to the tooth. Canine teeth are wanting, and the flat molars

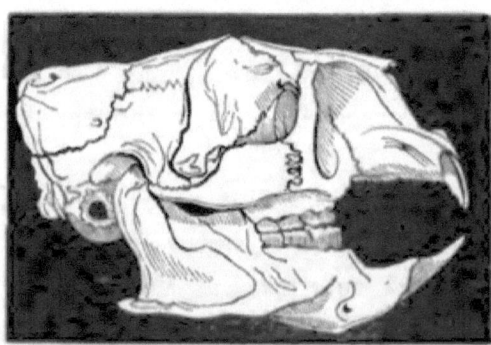

FIG. 165.—Skull of Rodent.

are separated from the incisors by a wide interval. (Fig. 165.) The hind legs of many Rodents, as the Hare and Jerboa, are much longer and stronger than the fore-legs. Most of the order are small in size, except the Capy-

bara, Beaver, and Porcupine. The order also contains
Squirrels, Rats, Mice, and Agoutis. The Beaver has a
smooth, unconvoluted brain, yet shows great ingenuity
in constructing its dwelling, felling logs with its teeth,
building them into a dam, and arranging others as a
shelter, plastering them with mud made into mortar by
its flat, trowel-like tail. The Flying-squirrel (*Pteromys*)
possesses a cloak of skin, stretching between the fore and
hind limbs, enabling it to sustain short flights in the air.

Order 5. *Insectivora*, (insect-eating.) These are di-
minutive animals, as the Shrew, the Hedgehog, and the
Mole. They have incisor, canine, and molar teeth, and
the latter have numerous pointed cusps. They have a
long muzzle, short legs, and clavicles. The feet are
formed for walking or grasping, and are plantigrade,
five-toed, and clawed. The Hedgehogs have a spiny
exoskeleton, covering the entire body, and lined by a
broad muscle, which, when it contracts, rolls the animal
into a ball.

Order 6. *Cheiroptera*, (*cheir*, a hand; *pteron*, a wing,)
are distinguished by long fore-limbs, adapted for flight,
the fingers being very long, and united by a membraneous web. The toes and one or two of the fingers are
armed with hooked nails. The Bat may be called the
only true flying Mammal, since it is capable of rapid
and long-continued flights. (Fig. 166.) The Vampire-
bat has a curious leaf-like expansion of the skin covering the nose. The ears of Bats are very large, and
copiously supplied with nerves of touch. The sense of
hearing is also acute.

Order 7. *Cetacea*, (*Ketos*, a whale,) are fish-like in form and habits. They are the largest of all living forms, and, next to the Elephant, have the heaviest brains.

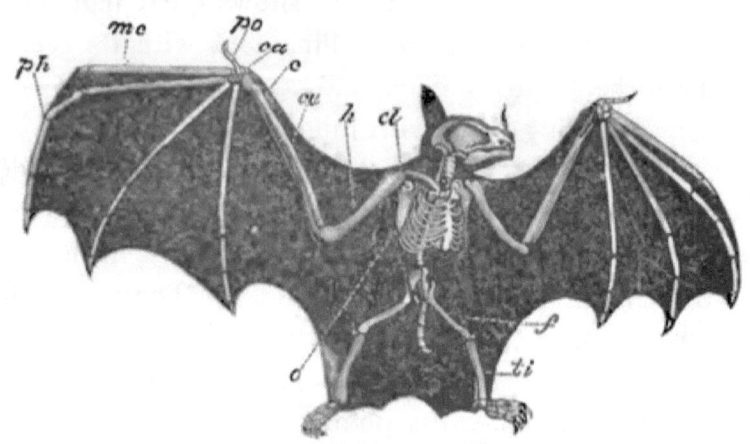

FIG. 166.—The Skeleton of Bat. *cl.* Clavicle. *h.* Humerus. *cu.* Ulna. *c.* Radius. *ca.* Carpus. *po.* Thumb. *mc.* Metacarpus. *ph.* Phalanges. *o.* Scapula. *f.* Femur. *ti.* Tibia.

The nostrils are on the top of the head, and constitute the *blow holes*, or *spiracles*. This order includes the Whales and Dolphins. All have a large horizontally flattened caudal fin. The head is large, often forming half the bulk of the animal. The Whalebone Whales (*Balænidæ*) are toothless, but in the Greenland Whale, the largest of the group, which sometimes attains a length of sixty or seventy feet, we find rudimentary teeth in the embryo. The Toothed Whales (*Odontoceti*) have many conical teeth in the lower jaw. The Sperm Whales are in this division. In them the head is large and abruptly truncated, and the nostrils are at the end of the muzzle. The *Delphinida*, comprising the Dolphins, Narwhals, and Porpoises have teeth in both jaws. Many Cetacea have very small organs of smell, and in the Dolphins and Porpoises they are wanting.

Order 8. *Sirenia* (*seiren*, a siren, or Mermaid) are like the Cetacea in shape, but are herbivorous, and frequent great rivers and estuaries. They have both a temporary and permanent set of teeth, a narrow brain, and nostrils on the top of a large snout. The Dugong and Manatee are illustrations of this order.

Order 9. *Proboscidia* include the Elephant, the extinct Mastodon, the Dinothere, and the Mammoth. There are no canine teeth, but the incisors are prolonged into *tusks*, which in the Elephant grow from the upper jaw, in the Dinothere from the lower jaw, and in the Mastodon from both jaws. The nose is prolonged into a long, flexible, sensitive trunk, which is terminated by a small prehensile appendage like a finger. Cuvier counted 20,000 distinct muscles in an Elephant's trunk. The limbs are massive, each with five toes incased in hoofs, and with a thick pad intervening between the toes and the ground.

Order 10. *Ungulata*, or Hoofed Quadrupeds, have four well-developed limbs, each having not more than four complete toes, and each toe being incased in a hoof. The leg is therefore for support and motion, and not for prehension. They have temporary and permanent sets of teeth. The *Odd-toed* Ungulates include the Horse, the Rhinoceros, and the Tapir. The Horse, which with the Ass and Zebra, made up the old order of Solidungula, has only a single perfect toe on each foot, coated with a nail, called a hoof, so that the horse walks and runs not merely on its toes, but on its nails. The Rhinoceros has three toes on each foot, and carries one or two horns on the skin of the nose.

The Tapir has four toes on its fore feet, and three on its hind feet, a short snout, projecting nasal bones, and a short tail.

The *Even-toed* Ungulates—the Hog, Hippopotamus, and Ruminants—have two or four toes. The Hog and Hippopotamus have the four kinds of teeth: incisors, canines, bicuspids, (or premolars,) and molars, and in the

FIG. 167.—Stag.

wild state are vegetarian. The Ruminants have two toes on each foot, enveloped in hoofs which face each other by a flat side, so as to appear like a single hoof split, or cloven. There may be two supplementary hoofs behind, but they do not usually touch the ground. All chew the cud and have a complicated stomach. They have incisors in the lower jaw only, and these are apparently eight; but the two outer ones are canines. With few exceptions, as the Camel, all Ruminants have horns, which are in pairs. Those of the Deer (Fig. 167) are solid, bony, and deciduous; those of the Giraffe and Antelope are solid, horny, and permanent; in the Goat, Sheep, and Ox they are hollow, horny, and permanent.

Order 11. *Carnivora*, or Beasts of Prey, have four long, acute, canine teeth, and there is a gap between the incisors and canines of the upper jaw for the reception of the lower canine. There are usually six incisors in each jaw. The digits always have sharp and pointed claws. The body is covered with hair.

The order is divided according to the peculiarities of the limbs. (Fig. 168.)

The *Pinnigrada* comprise the Seals and Walruses. The fore feet are webbed and form paddles. The hind feet are at the end of the body,

Fig. 168.—Toe of Lion. *a*, With the claw extended. *b, c,* Without the skin, retracted and extended.

enveloped in the integument, and in action they resemble the screw of a steam-ship. They live on fish.

The *Plantigrada* have the whole, or nearly the whole, of the hind foot in the form of a sole, which rests on the ground. The claws are not retractile; the ears are small, and tail short. Bears, Badgers, and Raccoons are well-known examples.

Digitigrada walk on the tips of the toes, and keep the heel raised above the ground. It includes the fierce and powerful Cats, Pole-cats, Ferrets, Weasels, Dogs, Hyænas, Jackals, Otters, etc. The Cats, (*Felidæ*,) embracing Lions, (Fig. 169,) Tigers, Leopards, Panthers, and Cats,

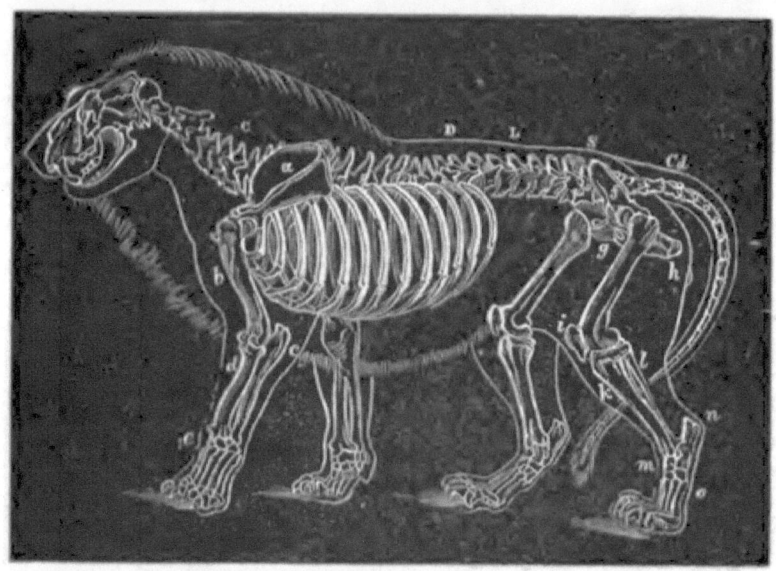

FIG. 169.—Skeleton of the Lion, (*Felis Leo*.) C. Cervical vertebræ. D. Dorsal vertebræ. L. Lumbar vertebræ. S. Sacral vertebræ. C. d. Caudal vertebræ. *a*. Scapula. *b*. Humerus. *c*. Ulna. *d*. Radius. *e*. Metacarpus and phalanges. *f*. Ilium. *g*. Femur. *h*. Ischium. *i*. Patella. *k*. Tibia. *l*. Fibula. *m*. Tarsus. *n*. Os Calcis. *o*. Metatarsus and phalanges.

have retractile claws, and the radius rotates freely on the ulna. They have also a prickly tongue.

Order 12. *Quadrumana* (four-handed) differ from all other Mammals by having each of their four limbs terminated by hands, in which the thumb is opposable to the other digits. (Fig. 170.)

The order is subdivided according to the position of the nostrils, into 1. Strepsirhines, or Monkeys with twisted nostrils, as the Lemurs and Aye-ayes, which

are the lowest of the monkey tribe.   2. Platyrrhines, or
Monkeys with simple sub-terminal nostrils, as the Spider-
monkeys.   These are South American, or New-World
monkeys, with prehensile tails.   The Howling-monkey
(*Mycetes*) has a curious modification of the larynx in the
shape of a bony drum attached to the hyoid bone, with
which it produces discordant shrieks.   3. Catarrhines, or

FIG. 170.—Quadrumana. Baboons.—1. Mandrill, (*Papio maimon*.) 2. Chacma, (*Chacma Porcarius*.) Monkeys.—3. Mona, (*Cercopithecus mona*.) 4. Howler, (*Mycetes*.)
5. Spider, (*Ateles*.)

monkeys with oblique nostrils, approximating below,
separating above, as the Gorilla and Chimpanzee. This
division includes the highest, or anthropoid Apes of the
Old World. They are all four-thumbed. The tail is not

24

prehensile, and is often quite rudimentary. The canine teeth are large. (Fig. 171.) The arms are long; in the Chimpanzee reaching to the middle of the tibia, when hanging down.

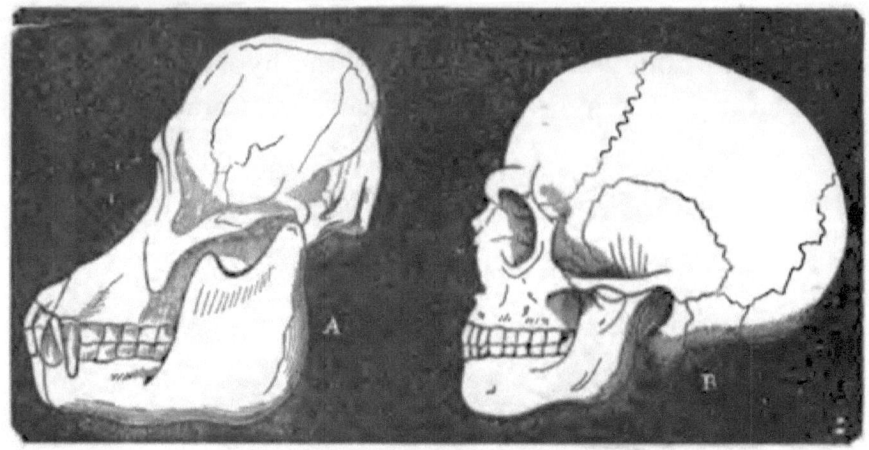

FIG. 171.—A. Skull of the Orang-outang. B. Skull of an adult European.

Order 13. *Bimana*, (two-handed,) contains but one genus and one species: *Homo*, or Man. (Fig. 172.)

Man differs from all animals in being an erect biped. The vertebrate type, which in all other cases is horizontal, in him is vertical. No other animal habitually stands erect; in no other are the fore-limbs used exclusively for head purposes, and the hind pair solely for locomotion. His limbs are parallel to the axis of his body, not perpendicular. They are nearly equal in length, but the arms are always a little shorter than the legs. In the Apes the arms reach below the knee.

Man only has a finished hand, which is a perfect organ of touch, and most versatile in movement. The foot is planted upon the ground by the entire length of the sole. The Gorilla has an inferior hand and an inferior foot.

The hand is clumsier and with a shorter thumb than man's, and the foot is prehensile, and is not applied flat to the ground.

Man is peculiar in his dentition. His teeth are vertical, of nearly uniform height, and close together. In every other animal the incisors and canines are more or less inclined, the canines project, and there are vacant spaces.

Man possesses two muscles (the *peroneus tertius* and *extensor primi internodii pollicis*) which are not found in the highest Apes. The origin of two other muscles is in Man altogether different from Apes. (The tibial origin for the *soleus*, and the calcaneal origin for the *flexor brevis digitorum*.)

The human skull has a smooth rounded outline, elevated in front, and devoid of crests. The cranium greatly predominates

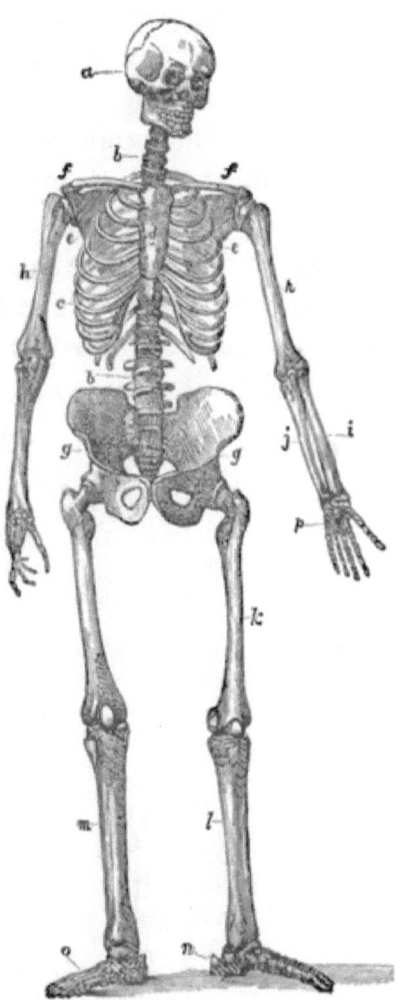

FIG. 172.—The Human Skeleton. *a*. Skull. *b. b.* Vertebral column, or Spine. *c*. Ribs. *d*. Sternum, or Chest bone. *e. e.* Scapulæ, or Blade bones. *f. f.* Clavicles, or Collar bones. *g. g.* Pelvic, or Hip bones. *h. h.* Humeri, or Arm bones. *i.* Radius, and *j.* Ulna, bones of fore-arm. *k*. Femur, or Thigh bone. *l*. Tibia, or Large bone of leg. *m*. Fibula, or Small bone of leg. *n*. Calcaneum, or Heel bone. *o*. Tarsal bones, or Bones of the foot. *p*. Carpal bones, or Bones of the wrists.

over the face, being four to one. Man differs from all
Apes in the absolute size of brain, and in the greater
complexity and less symmetrical arrangement of its convolutions. The brain of the Gorilla scarcely amounts
to one third in volume or one half in weight of that of
Man.

From purely morphological reasons, therefore, Man is
entitled to rank as a distinct order of Vertebrates. Other
considerations, to be referred to in the next chapter,
show that he should be regarded as a distinct type.

CHAPTER XVI.

THE HUMAN TYPE.

> The master-work, the end
> Of all yet done ; a creature, who, not prone
> And brute as other creatures, but endued
> With sanctity of reason, might erect
> His stature, and upright with front serene
> Govern the rest, self-knowing ; and from thence
> Magnanimous, to correspond with Heaven,
> But grateful to acknowledge whence his good
> Descends; thither with heart, and voice, and eyes
> Directed in devotion, to adore
> And worship God Supreme, who made him chief
> Of all his works.—MILTON.

1. IN the rapid panoramic survey of living forms, which is all our limits will allow, we have mainly confined ourselves to structural forms, barely glancing at the instinctive peculiarities which determine these forms for special ends. It is necessary to supplement our review by a reference to functions and endowments which the structure itself may not always indicate.

2. Biology includes not only Anatomy and Physiology, but Psychology also. "The naturalist studies the instincts of the Ants and the Bees. When he attempts the history of Man, shall he put aside that which in him represents these instincts? Evidently not. Consequently he must not stop with the body. He must consider the intelligence which is in us, and which, up to a certain point, we have in common with animals; he must show that it is this element of our being which recog-

24*

nizes the outer world, which judges, which aspires. His work will be very imperfect if he neglects this something of which the nature escapes us, but of which the power is such, that through it man has not only vanquished all animals, whatever their defenses, their size, or their strength, but he has overcome, and made to work as his servants, even the immutable forces of the inanimate world." *

3. We have seen that the lower animals partake of living structures and organs, the same in essential character and objects as those of man. Careful observation will show that they also possess many mental or psychological endowments, such as we find in the human type. The differences of animated nature are differences of degree rather than of essential nature. So far as we can see, all animals have self-consciousness and volition, and many exhibit unmistakable signs of reason.

4. On page 283 is an outline plan of the psychical endowments of man, with the objects constantly influencing him and the normal activities of his being. It begins with the most general and elementary properties of animal life, and rises to the highest special powers of human nature. More than an outline cannot be attempted, since an elaborate exposition would require a large volume.

5. It will be seen that we have given prominence to consciousness in the plan referred to. This is because it is an essential condition of every mental operation. It is the knowledge which the mind has of its own operations.

* Quatrefage's "Natural History of Man."

Objective.	SUBJECT.	*Subjective.*

Spiritual and Rational Objects, (the Good, Beautiful, True, etc.)

Consciousness of Spiritual Beings and Principles.

Will.

Judgment.

Faith.

Conscience.

The Mind itself.

Consciousness of Ideas, Sentiments, Emotions, Imaginations.

Fancy.

Thought.

Memory.

Objects of Sense.

Consciousness of Sensations.

Perception.

Voluntary Motion.

Obscure Ideas.

Organic Sensibility.

Consciousness of States of Body.

Instincts, or Consensual Actions.

Consciousness of the Body—or of Self.

The Organic Life.

Involuntary Motion.

Afferent Impressions, interrupted by Disease or Depravity.

Efferent Actions, deranged by Sleep, Intoxication, Insanity, and Disease.

In the general account of the nervous system (Chap. XV., Sec. 1) it was stated that many motions were merely reflex and involuntary. Many such motions are also without consciousness. It is probable that a very large proportion of the movements of the lower animals are of this character. Other motions depend on organic contractility responsive to an external stimulus, as when a piece of muscular fiber contracts on being scratched with a pin. Ciliary motions, the closure of the leaves of Venus'-flytrap (*Dionæa*) on being touched by an insect, and the movements of the Sensitive plant, may thus be accounted for. Some motions, as the sleep of plants, depend on the periodicity of functional activities, and others, as the bursting of seed-vessels, may be owing to Endosmose. Mere movement, therefore, is far from indicating consciousness.

" How early does consciousness arise ? If we interpret, as we are constantly doing, the experience of lower animals by that of higher ones, we should answer, With the very commencement of animal life. Indeed, nothing but conventional sentiment would prevent our attributing, under this method, a feeble consciousness to some plants. If, however, we reason from the character of the nervous system, which is undoubtedly the sole organ of consciousness, and from the stages in development at which a conscious experience can enter as a profitable factor, we shall be inclined to believe that consciousness especially characterizes the Vertebrata, and appears first in the higher Articulata and Mollusca. The phenomena of consciousness undoubtedly increase greatly in vigor and in value as we pass up through the Vertebrata, and this form of activity is, in its governing relations, collected and specialized in the cerebrum." *

Without attempting to dogmatize upon a subject so imperfectly known, we may suggest that many of the habits of Ants, Bees, and even of animals of a more primitive type, afford as good evidence of consciousness as the actions of human creatures themselves.

6. The consciousness of self, or general corporeal sensitiveness, is the earliest sign of individuality, or personal knowledge. This is previous to the senses, and independent of the nervous system. It manifests itself in animals without nerves, as the Polyps, and seems to be a necessary attribute of animal life. Yet this most primitive and most clearly innate faculty implies mind, for by it we know that our body is *our* body. The corporeal structure is an object of which the mind takes cognizance. The presence of this sensitivity proves the existence of something distinct from the body.

7. The consciousness of the physical conditions or states of the body—as tonicity, languor, hunger, thirst, warmth, and cold—has been termed common sensation, or *cænæsthesis*. It is especially conducted, at least in the higher animals, by the ganglionic or sympathetic system of nerves. By means of the connection of this with the cerebro-spinal system the various affections of the mind and body mutually act upon each other, rendering the phenomena quite complex. Certain obscure ideas, of which one may be said to be half-conscious, and which taken together make up what we call the disposition or temper of a man, are the result of organic sensibility act-

* Bascom's " Comparative Psychology."

ing upon the common sensation. What Dr. Carpenter terms " consensual actions " may also originate here, as well as from sensation proper. In this term that eminent physiologist includes all the purely instinctive actions of the lower animals, which make up, with the " reflex," nearly all the animal functions in many tribes, and which are peculiarly elaborate in their character and wonderful in their results in Insects. Such automatic and involuntary actions as vomiting excited by the sight of a loathsome object, a bad smell, or a disagreeable taste, or laughter excited by tickling, are also classed under this term.

8. Sensation, or special sense, is caused by an impression on certain parts of the nervous system, which are hence called sensitive. For sensation two things are necessary: an impressible state of the sensitive organs, and a perception by the mind.

9. Perception is the evidence we have of external objects by means of the senses. It is necessary that the organs and nerves be sound, or false perceptions will result. Ringing noises in the ears, floating specks before the eyes, and many spectral illusions, have their origin in a diseased condition of the organs. Yet that perception is an attribute of mind is evident from the fact that attention is required. The senses may be impressed by their appropriate objects, but without attention they are not perceived.

10. Memory implies a former conscious experience, either of a physical or mental kind ; its retention, revival, and recognition. The laws of memory, as they are called, or circumstances which excite recollection, have been

enumerated, as resemblance, contiguity, cause, effect, and contrast.

11. The mind itself may produce in the sphere of consciousness, ideas, sentiments, emotions, and imaginations. For the manifestation of mental phenomena it is doubtless important to have continuously healthy nerve-structure and other bodily organs, since the most ac complished artisan cannot exhibit his full powers with imperfect tools and materials; yet as the injury or destruction of the implement is no proof of the annihilation of the artisan, so the injury or even destruction of the body may not affect the soul. The mind is popularly supposed to be dependent on the brain, yet medical authorities show that every portion of the brain has been, in one instance or another, destroyed or disorganized without affecting what are supposed to be the corresponding intellectual powers. Abercrombie tells of a lady in whom one half the brain was disorganized, but who retained all her faculties to the last, and many such instances are on record. There is no constant relation between the integrity of mind and body. The mind may suffer intense agony while the body is in perfect health, or remain in calm serenity while the body is tortured or is losing its vital powers.

12. Ideas, in a general sense, refer to any thing present to the mind as an object of thought, whether present really or representatively. Some ideas are related to experience, as the principles of mathematics, notions of figure, extension, number, time, and space. Others are independent of sensible representation, as the ideas of good and evil, just and unjust, true and false.

13. Sentiments refer to feelings of esteem, gratitude, patriotism, etc., but emotions to mental pleasure or pain. The emotions are often very complex, and influence every part of the nature, physical and mental; as hope, joy, melancholy, love, and anger.

14. Imagination is a term which represents the power which the mind has of combining ideas. The images produced by this faculty are sometimes so vivid as to affect the organs of sense, and occasion morbid sensual delusions, as well as to influence the organs of motion, secretion, etc. No proof could be more positive of the independent agency of the mind. In its highest degree imagination leads to creative fancy, or poetic power. In some of its flights it may encroach upon the prerogative of conscience, and lead to self-deception unless held in check by the precepts of Divine revelation.

15. Conscience has been called the moral sense, moral faculty, moral judgment, and susceptibility of moral emotions. It may also be termed the inspirational capacity of the soul. It is that faculty, or combination of faculties, by means of which we have ideas of right and wrong respecting actions, and corresponding feelings of approbation or disapprobation. Faith, in the scriptural sense of the term, is not belief, but the volitional activity of the mind in the sphere of the conscience.

16. Judgment is the decision of the mind after comparison. It is altogether a mental function. It is an act of the mind upon and within itself.

17. Volition is the dominion exercised by the mind over itself, employing or withholding its faculties in any particular action. It is synonymous with free agency,

and is an essential attribute of spirit, since the very idea
of spirit supposes self-action. Feuchtersleben judiciously
distinguishes between the essential freedom of the spirit
and the freedom of the spirit linked to the body. He
shows that freedom may, first, limit itself, so far as the
spirit makes itself the slave of sin or error; second, it may
be limited by physical laws; third, it may be limited by
organization. As to the first, the free man is good and
wise; as to the second, powerful; as to the third, healthy.

18. This brief examination of human endowments
shows as great a difference between men and brutes as
exists between animals and vegetables, or between vegetables and the mineral world. It is considered by many
that each department of nature becomes higher through
the addition of something which the next below it did
not possess, and as the differences of the animal and
vegetable world form successive additions to a common
original plan or system of organization, we find foreshadowings or prophecies of the characteristics of higher
forms. Thus the regularity of the crystal suggests to
the imagination the organization of the plant, and the
motions of plants foreshadow the nervous system. Thus,
too, the higher animals have vague and indistinct analogies of the vast endowments of man.

19. The unity of man was generally conceded by the
early naturalists, but has been largely debated in recent
times. Agassiz himself held to different creations, although believing they were a unit as to intellectual and
moral nature. The discussion continued, until a few
years ago it appeared to be the settled creed of men of
"advanced" views to deny man's unity. Yet one point
25

after another has been changed, until, in the language
of Mr. Tylor,* "it may be asserted that the doctrine of
the unity of mankind now stands on a firmer basis than
in any previous ages."

20. Respecting the antiquity of man upon the earth,
it is very plain that the differences between the Hebrew,
Samaritan, and Greek Pentateuch are such as to forbid
any settlement of the question by a reference to the
Scriptures. Long before the modern discussions on this
subject biblical scholars doubted if it was the design of
the Scriptures to reveal either the antiquity of man or
the age of the earth. Yet the discovery of human remains at Abbeville and other places, the remains of lake-
dwellings in Switzerland, and the shell heaps in Den-
mark, are nowise inconsistent with the view of a
degradation of some races from a more highly civilized
condition. The ruins of ancient nations certainly point
to an early civilization which was remarkable for extent
and splendor. As to the time required for these changes,
Dana, in his " Manual of Geology," says: " The evidence,
as it at present stands, does not necessitate the carrying
of man back in past time, so much as the bringing forward of the extinct animals toward our own time."

21. The numerous varieties of the human species may
be divided into four principal races, which comprise secondary and mixed races, each including a number of
families and nations: 1st. The White race, also, but er-
roneously, called Caucasian. Its original country, judging from the comparison of languages and historic testi-
mony, lay between the Mediterranean, the Red Sea, the

* Art., Anthropology, in Encyc. Brit., ninth edition.

Indian Ocean, the steppes of Central Asia, and the Himalaya Mountains. From thence it has spread into India, Arabia, Syria, Asia Minor, and Egypt. 2d. The Red, inhabiting only America. 3d. The Yellow, which has existed in China from remote antiquity, and has spread into all countries inhabited by Mongolians. 4th. The Black, which belongs to Central and Western Africa, and is distributed over the tropics from the east coast to Australia. It is doubtful if either of these races represents the primitive type of man.

22. We close our brief survey of life with the religious sentiments of the Psalmist: "I will praise thee; for I am fearfully and wonderfully made: marvelous are thy works; and that my soul knoweth right well. My substance was not hid from thee, when I was made in secret, and curiously wrought in the lowest parts of the earth. Thine eyes did see my substance, yet being imperfect; and in thy book all my members were written, which in continuance were fashioned, when as yet there was none of them. How precious also are thy thoughts unto me, O God! how great is the sum of them! If I should count them, they are more in number than the sand: when I awake, I am still with thee. .... Search me, O God, and know my heart; try me, and know my thoughts; and see if there be any wicked way in me. and lead me in the way everlasting."

INDEX.

THE END.